TIMBER RATTLESNAKES IN VERMONT AND NEW YORK

T0317029

Timber Rattlesnakes

IN VERMONT AND
NEW YORK

BIOLOGY, HISTORY, AND THE
FATE OF AN ENDANGERED
SPECIES JON FURMAN

UNIVERSITY PRESS OF NEW ENGLAND

University Press of New England
An imprint of Brandeis University Press

© 2007 by University Press of New England
All rights reserved

For permission to reproduce any of the material in this book, contact
Brandeis University Press, 415 South Street, Waltham, MA 02453
or visit brandeisuniversitypress.com

Ebook ISBN: 978-1-61168-816-0

Printed in the United States of America

Library of Congress Cataloging-in-Publication Data
Furman, Jon.
Timber rattlesnakes in Vermont and New York : biology, history, and the fate
of an endangered species / Jon Furman.
    p.   cm.
Includes bibliographical references and index.
ISBN-13: 978-1-58465-656-2 (paperback : alk. paper)
ISBN-10: 1-58465-656-5 (pbk. : alk. paper)
1. Timber rattlesnake—Vermont.   2. Timber rattlesnake—New York (State)
I. Title.
QL666.O69F87 2007
597.96'3809743—dc22   2007027450

CONTENTS

ILLUSTRATIONS

ACKNOWLEDGMENTS

There are many people who contributed in one way or another to the writing of this book. I am eternally grateful to each and every one of them. I am especially grateful for the unwavering support and guidance of my wife, Kitty, and my editor, Ellen Wicklum, since the beginning. I would also like to thank Bill Brown, Marty Martin, Randy Stechert, and Dan Keyler for their tremendous support and the wealth of knowledge and experience they were willing to share with me. To a large extent I owe my own expertise on the timber rattlesnake to theirs. Finally, I'll always appreciate Lester and Carol Reed, Bill Galick, and Art Moore for having introduced me to the world of timber rattlesnake bounty hunting.

INTRODUCTION

Probably every grade school in rural, small-town America in my childhood, during the 1950s and 1960s, had a boy who was crazy about amphibians and reptiles. I know our school in central New Jersey had one. In my case I wasn't that boy, but one of my best friends was. John regularly caught and kept a variety of snakes at his home in Somerset County. Once I was with him when he captured a large black racer (*Coluber constrictor*). It was an extremely feisty snake and struck him twice on the leg, drawing blood, but John didn't panic in the least. For years afterwards I admired him for his ability to stay cool under attack and capture that snake without hurting it.

In June one year while we were fishing a small trout stream on Schooley's Mountain in Long Valley, New Jersey, we shared a brief but electrifying snake experience. We had fished the stream up to its source, a large spring-fed pond near the top of the mountain. The outlet from the pond had a concrete dam and just downstream from its spillway was a man-made, stone embankment about twenty yards long and twenty or so feet high, along one side of the stream. I was casting down below the spillway with a roar of water noise around me, and the last thing on earth I was thinking about was danger in any form.

I had no idea where John was since I hadn't seen or spoken with him for five minutes or longer. It turned out that he was directly above me on top of the stone embankment on a flat area of smaller rocks and stones probably six feet wide. Suddenly he yelled out, "Copperhead!" (*Agkistrodon contortrix*). There was more than excitement in his voice. There was warning. Barreling down the rocks directly at me was a powerful, fast-moving snake. I instinctively stepped out of its way as it passed within inches of my feet. Even though it was trying to get to cover as quickly as possible and was not interested in me in the least, my heart was pounding. Years later I can still feel the excitement of the moment, and John's excitement in particular. "I came right on him while he was sunning," he explained. "I could see his head clearly. What a beauty!" Coming face to face with one's first venomous snake in the wild can have that effect on a person.

It was almost forty years later that I had my second encounter with a venomous snake in the wild. It took place in upstate New York near the Vermont border, approximately two hundred and fifty miles north of Long Valley, and this time the snake was a four-foot-long, male timber rattlesnake (*Crotalus horridus*). This encounter was every bit as exciting as seeing the copperhead in Long Valley in my boyhood. In a matter of seconds, it ignited an interest in timber rattlesnakes that will undoubtedly remain with me for the rest of my life. People often ask me how I became involved with these snakes in the first place. More than anything else it was my curiosity.

After moving to Rutland County, Vermont, in the early 1970s, I began to hear stories, some obviously exaggerated for effect and others totally plausible, about timber rattlesnakes in the western part of the county. After hearing these stories for nearly thirty years, I had finally heard enough of them to propel me into a fact-finding mission, no matter where it might take me. In the winter of 2002 I decided the time had finally come to find out the truth about the rattlesnakes of Rutland County, and I soon found myself on a remarkable journey in which I met and became friends with a number of nationally known timber rattlesnake experts, interviewed and came to know a number of former timber rattlesnake bounty hunters, and observed scores of timber rattlers in the wild. I also learned a great deal about their current status in Vermont and New York, their past and present range, their habitat and their biology, as well as a fascinating period in Vermont and upstate New York history in which these snakes were hunted for a bounty in four adjacent counties.

When I look back at everything I have learned about the timber rattlesnake, it's difficult to single out the most important element that inspired me to write this book, because there were several: the biology of the species, man's relationship to the species in the northeast dating back to pre-colonial times, and my fascination with the timber rattlesnakes of Rutland County, Vermont, and Warren, Washington, and Essex counties across the border in New York. I find it remarkable that the timber rattlers in these four counties are in the uppermost reaches of their range in the northeast, and that they were severely depleted by, but were able to survive, a long period of bounty hunting that lasted from the 1890s until the early 1970s. I was inspired to document the history of this bounty hunting, the impact it had on the snakes, and what's been done to help prevent the snakes from becoming extinct since the bounties on them were discontinued over thirty years ago. I was also inspired to discuss what the future may hold for the timber rattlesnakes in Vermont and New York, and particularly the "border rattlers" of the four former bounty hunting counties.

The reader may wonder why I haven't revealed the names of towns or mountains in the former bounty hunting counties on the Vermont/New York border, and why I have largely refrained from mentioning the names of counties in southern New York when discussing specific timber rattlesnake locations. There are a number of reasons for these omissions. Local people are usually aware of rattlesnakes wherever they occur, but it's not "locals" that concern me. It's the person who is looking for remote dens to remove snakes from, for sale to collectors, or to kill for sport, or to keep for use in one way or another. Unfortunately, there are people who would think nothing of crossing state lines and breaking the laws that protect these snakes in Vermont and New York, and other states in the northeast, if they thought they could get away with it. There are also others who simply want to see, and possibly photograph, these reptiles in the wild. Their intentions are seemingly innocent and harmless, but timber rattlesnakes have a relatively low tolerance for human disturbance. If enough people tramp around on den sites and basking areas, these snakes may alter patterns of behavior that they have followed for thousands of years. When this happens, they become harder and harder to find, and therefore harder for researchers to study.

TIMBER RATTLESNAKES IN VERMONT AND NEW YORK

CHAPTER 1 : GETTING TO KNOW THEM

PHYSICAL TRAITS AND CHARACTERISTICS

During my initial encounters with timber rattlesnakes in the wild, one of the first things I realized about them is how well they can blend into the rocky, leaf-strewn forests they inhabit, and how difficult they can be for a first-timer in their habitat to detect. Nature has provided them with a wide variety of colors and patterns that give them an extraordinary camouflaging effect, often referred to as cryptic coloration. Throughout the major portion of their range, which includes most of the states to the east of the Mississippi River and a few to the west of the river as well (see pl. 10), timber rattlesnakes fall into two different color variations: yellow morphs or black morphs. Yellow morphs typically have a background color of yellow or tan in association with gray or dark brown dorsal and lateral blotches that join to form bands. Their heads are invariably yellow or tan. Their bellies are a yellowish cream color, lightly flecked with either black or brown, and their tails are uniformly black (see pl. 5). Black morphs tend to have a background color of black or gray in association with black or dark gray blotches. Their heads are black or gray. Their bellies are white, or cream, with either gray or light yellow mottling, and their tails are uniformly black (see pl. 3). Therefore, in terms of their cryptic coloration, the only thing that the two morphs have in common is the color of their tails.[1]

Another color variation is seen in timber rattlesnakes in the southeastern and southern portion of their range. This is the southern morph variation, which occurs in what are often called canebrake rattlesnakes. Canebrakes have a background color that is usually a light shade of gray, pinkish gray, or pinkish buff, while their blotches are similar to those of yellow morphs and black morphs. Where yellow and black morph coloration is ideally suited for the deciduous forests that predominate throughout much of the timber rattlesnake's range in the east, the coloration that has evolved in canebrakes is probably more suited for the coastal and lowland regions they inhabit in the southeastern and Gulf Coast states.[2]

A fourth color variation, very similar to the southern morph, is seen on timber rattlesnakes west of the Mississippi River from the Ozarks northward. This variation is referred to as the western morph.

The one biological trait that sets rattlesnakes apart from all other snakes, venomous and nonvenomous alike, is their rattles. With the exception of several species of rattlesnakes inhabiting the islands of Baja California, all rattlesnakes have well-defined rattles. Before experiencing the "rattling" of my first rattlesnake in the wild, I had heard from an old Vermont timber rattlesnake bounty hunter that there is nothing like hearing the buzz of an agitated timber rattler that's in close proximity, especially when you haven't yet seen the snake. "It can bring you to your senses real fast," he claimed. He was right. I suppose it's possible for someone who hears rattling over and over again in laboratory situations to develop some degree of "immunity" or indifference to the sound, but I can't imagine ever getting to the point myself where the buzz of an unseen, nearby timber rattler in the wild won't give me an instantaneous surge of fear and excitement. That's basically what rattling is supposed to do to a would-be predator.

No one knows how rattles evolved. They are an intricate overlapping series of multi-lobed, hollow segments that articulate loosely enough with each other to allow for movement, while a built-in mechanism called a gripping claw prevents them from falling apart. It's the movement of these segments against each other that produces the well-known sound of rattling. A popular theory on the origin of the rattle was popularized by Samuel Garman, a Harvard University herpetologist and ichthyologist in the late 1800s. According to Laurence Klauber, perhaps the best known rattlesnake authority of all time, Garman believed that the rattle developed as a specialization of the enlargement of the tail cone that terminates the tail of most of the world's snakes. The difference between rattlesnakes and all the other snakes with tail cones is that rattlesnake tail cones are larger. In Garman's theory, the rattle evolved as a result of the skin that hung up on these enlarged cones every time the ancestral rattlesnakes shed their skins. In short, the skin that built up in many consecutive sheddings would harden and thicken their tail cones, thus becoming the all-important first step in the evolution of the rattle as we know it today.[3]

Where there are plausible explanations of the evolution of the rattle, in my opinion this is not the case with the purpose of the rattle. Laurence Klauber devoted over ten pages to this subject in his renowned 1956 monograph on rattlesnakes entitled *Rattlesnakes*.[4] The most widely accepted explanation regarding the purpose of the rattle is commonly referred to as the bison theory.[5] Its premise is that the rattle may have evolved to give the earliest rattlesnakes a defense mechanism against being stepped on and crushed by large land mammals such as bison that roamed the plains and open grasslands of what is now Mexico and the southwestern United States millions of years ago. Those who ascribe to this theory believe that any large beast such as a bison would experience enough pain after being bitten by one of these reptiles to make it try to avoid the buzzing rattle of an agitated rattlesnake in the future. Inherent in this theory is the premise that other unbitten bison would learn to avoid rattling as well.

Two far less believable theories are that the rattle developed as a way for rattlesnakes to locate mates, or to warn other rattlesnakes of impending danger. Klauber details a number of reasons and personal experiences for why these theories don't hold water, but one reason stands above all others. Rattlesnakes don't have external ears and, as a consequence, cannot detect high-frequency airborne sounds of the rattle. Klauber was aware of this fact as a result of experiments he conducted during his thirty-five-year tenure as a Consulting Curator of Reptiles at the San Diego Zoo.

Although rattlesnakes are the only snakes that have the ability to make a rattling noise with a highly evolved rattle, there are other species of snakes that can make a noise quite similar to the sound of rattling by vibrating their tails rapidly in dry leaves.[6] In the northeastern portion of the timber rattlesnake's range a number of snakes are known to exhibit this behavior and can therefore be misidentified as timber rattlers. In this group are copperheads, eastern milk snakes (*Lampropeltis triangulum*), black racers, and black rat snakes (*Elaphe obsoleta*). When hearing a rattling sound in association with a coiled snake, many people are convinced that they are in the presence of a timber rattlesnake, especially if they are in timber rattlesnake country. Game wardens in Vermont have told me that they periodically receive calls from distraught land owners who believe that a timber rattlesnake is on their property. When the wardens follow up on these calls, very often they discover a totally harmless eastern milk snake.

Timber rattlesnakes gain a new rattle segment every time they shed except for the first time, about ten days after they're born, when all they gain is what's referred to as a button. Dr. William S. Brown, one of the top timber rattlesnake

FIGURE 1. William S. Brown reflecting on a rattlesnake den in Washington County, New York. Photo taken on 10/10/2003 by Jon Furman.

researchers and authorities in the country, explains that it's possible to age these snakes fairly accurately by dividing the number of segments in their rattles by the average number of sheddings per year in their area. This can work if a snake has retained its button and each subsequent segment of its rattle. What makes it difficult is the fact that timber rattlesnakes frequently lose segments off their rattles in the wild, and it's rare to find an adult snake with its rattle, including its button, intact. When examining a rattle without a button one can only give a rough estimate of a timber rattlesnake's age. One thing that helps is the fact that timber rattlesnake rattles tend to taper up through a snake's tenth shedding but afterward this tendency discontinues.[7]

Brown has seen at least one exceptional rattle in his career. It had twenty segments and was on a large male timber rattler that he captured in his study area in northeastern New York around 1981. There was a tone of reverence in his voice when he first told me about the snake. Apparently he has never seen another snake like it and has more or less given up hope that he ever will again. I asked him if it could have been a mutation, to which he replied, "No, but it was special." In over twenty-five years in the field Brown has seen rattles with fourteen or more segments in fewer than 1% of the hundreds of adult male and female snakes he has captured and released.[8]

I feel fortunate to have seen a large male timber rattler with a long, seventeen-segment rattle on a field trip with Randy Stechert, one of the top timber rattlesnake field biologists in New York State, in the summer of 2003. The large black morph snake was one that Randy had surgically implanted with a tiny transmitter, had been tracking with a receiver, or radio tracking, in southern New York for several years, and was very familiar with. The first thing I noticed as we came upon the snake simultaneously was the size of its rattle. I knew right away that it was unusually long and that I might not ever see one like it again. It was well over two inches in length and stood out so prominently that it could easily be detected on the long, stretched-out snake from over ten feet away.

Something that concerned me as I first began to learn about the timber rattlesnakes of Vermont and northeastern New York, by interviewing some of the old rattlesnake bounty hunters who had hunted in the region, was that they weren't always in accord on various biological facts. One area in which this became evident centers on the shedding tendencies of the snakes. Lester Reed and his daughter, Carol, who were well known Vermont rattlesnake bounty hunters and producers of timber rattlesnake oil, always thought the snakes from the den they hunted in Rutland County shed their skins once a year. They noticed that the snakes tended to be darker in the spring and much more colorful in the fall, and they attributed this to their shedding at some point during the summer. Bill Galick, the legendary timber rattlesnake bounty hunter from western Rutland County, told me that the snakes shed two to three times a year depending on how much they eat. Art Moore, who was the best known and most prolific timber rattlesnake bounty hunter in the history of New York State, told me in one of my first interviews with him that the timber rattlers he kept in captivity shed twice a year, and he assumed that the same thing happened in nature.

W. S. Brown's ongoing capture-mark-and-recapture field study in the eastern foothills of the Adirondack Mountains in New York provided a lot of clarification on the subject. By clipping, using a pair of dissecting scissors, a recognizable and permanent number in the belly scales of all the timber rattlesnakes he captures, Brown has been able to track hundreds of individual snakes over the years and acquire a great deal of information about them.[9] Every time he recaptures a snake he has previously captured, marked, and released, a feat that has taken place over two thousand times in his career, he is able to learn a lot more about it than just its location. Among the things he learns is how much it has grown, whether females have given or are about to give birth, how old it is, what den it's from, etc.

One of the first things he ascertains is whether the snake is in what's known as the pre-shed mode, with its eyes filmed over and the color of its skin muted, or whether it has the intense coloration typical in post-shed snakes. After numerous recaptures of the same snakes, Brown has been able to come up with an average shedding rate of about 1.4 sheds a year for males and 1.2 sheds per year for females in his study area. There's a fair amount of variation in annual shed rates throughout the timber rattlesnake's range, but Brown's averages can be considered as an accurate standard to go by in Vermont and northeastern New York where the snakes' active season is about five months annually.[10]

As a rule, timber rattlesnakes in the southern portions of their range shed more frequently than their northern counterparts. This is because they're apt to have longer active seasons (eight to nine months in the case of canebrakes, or southern variant timber rattlesnakes) in which to forage. The more they eat, the faster they grow, and the more often they shed. W. H. Martin, another widely known timber rattlesnake researcher and authority, has learned that timber rattlesnakes in the southeastern coastal region, for example, have as many as four shedding events in their first full year, including the postnatal shedding of their pre-buttons within their first two weeks of life. After their first year, they average close to two sheddings annually. By the time a male timber rattlesnake in this region has gone through six full years, it has typically had fourteen shedding events and has shed at a rate of approximately 2.3 times a year. Female timbers in this same area will normally have thirteen shedding events in the same period of time, shedding at a rate of 2.1 times annually.[11]

Being farther south in the timber rattlesnake's range doesn't necessarily go hand-in-hand with a higher annual rate of shedding, however. Elevation can play a critical role in the equation. A prime example of this occurs in Martin's study area in the Allegheny Mountains on the Virginia/West Virginia border. He has observed that most of the rattlesnakes on what he refers to as the Allegheny Front tend to den at elevations over 3,500 feet, and shed at the rate of once a year after their eighth year.[12] This is a clear-cut example of elevation's impact on annual shed rates. Regardless of the fact that they are hundreds of miles farther south than the timber rattlesnakes on the New York/Vermont border, which typically den at elevations under 1,300 feet, they have a shorter active season and subsequently shed at a lower annual rate.

In addition to being venomous, or having the biological ability to produce and deliver a powerful poison, timber rattlesnakes are vipers, which puts them in company with about a third of the world's venomous snakes including many well-known ones such as the fer-de-lance (Bothrops lanceolatus) in South America and the gaboon viper (Bitis gabonica) in east Africa. No matter where they occur, all vipers have long, curved, hollow, folding fangs in their upper jaw, which allow them to inject their venom with great efficiency into whatever form of life they happen to bite. Their fangs have been compared to hypodermic needles, and rightly so. The rest of the world's venomous snakes, including elapids and certain colubrids, get by with a different fang design. Unlike their viper counterparts, their fangs are shorter and largely fixed, and almost totally devoid of any movement. This doesn't mean that snakes with this delivery system are less dangerous in their bite. In fact, the opposite is often the case. Most of the deadliest snakes on earth, including the king cobra (Ophiophagus hannah), black mamba (Dendroaspis polylepis), and taipan (Oxyuranus sp.) are not vipers. What they lack in fang design, they make up for with very toxic venom.

Perhaps the most interesting physical trait of timber rattlesnakes, other than their camouflage, their rattles, and their deadly fangs, is the pit that lies between their nostrils and eyes and classifies them as New World pitvipers. Not all vipers have pits, but those that do are able to distinguish, through the membranes in their pits, subtle variations in the infrared radiation emanating from objects that are in their immediate vicinity. This ability is significant to the survival of timber rattlesnakes as well as many other species of pitvipers in a number of ways. First and foremost, it greatly improves the quality of their vision during the hours of darkness. Their eyes are well adapted to seeing at night, but by combining something akin to thermal images of objects up to two feet away with their normal visual capacity, they are able to see the area immediately in front of them with great detail even during the blackest of nights.

According to Dr. Rulon Clark, a timber rattlesnake researcher at Cornell University, these snakes derive several significant advantages from their ability to detect infrared radiation. The defensive advantage is that it can help them home in on would-be predators that come dangerously close to them at night. The offensive advantage they gain is even more significant in Clark's opinion. There's no question in his mind that the timber rattler's super sensitivity to infrared radiation gives the species the ability to home in on, strike, and follow up on its prey at night. It also gives these snakes the ability to determine the warmest areas in their immediate vicinity, which enables them to better thermoregulate themselves during the day as well as at night by selecting warm rather than cool retreat sites.[13]

Timber rattlesnakes are formidable predators equipped with so many hunting advantages that it's difficult to place one above another in terms of its importance. One of these advantages, certainly, is their ability to blend into their environment. Even when timber rattlers are coiled up on an open forest floor with nothing but dry leaves around and under them, a person could walk within two to three feet of them and never notice them unless they moved or rattled. It's even harder to detect them when they're coiled up next to an object such as a fallen tree, a rock, or any form of vegetation that would tend to divert one's visual attention away from them. On more than one occasion I have been within two feet of these snakes and not been aware of their presence until the person I was with alerted me to them, or the snakes started to rattle. If it's that difficult for a person to detect these animals in their natural surroundings, it's very likely that many, if not all, of the animals that they hunt have trouble seeing them too.

If I had to pick the two greatest hunting advantages of timber rattlesnakes, I would go with their camouflage and the potent killing power of their venom. When a timber rattlesnake envenomates (injects venom into) any one of the wide variety of rodents upon which it preys, the end result is pretty much a given. Typically, the animal runs off a short distance, falls down in a semi-paralyzed state, kicks spasmodically, and dies in a matter of seconds from massive heart failure. The odds of the animal escaping without the snake's being able to find it are extremely low, and there is virtually no chance that it could survive, even if it managed to escape.

As well as timber rattlesnakes are able to conceal themselves and disable and kill their prey, they could go for very long periods of time without eating if it weren't for a few additional hunting advantages. Clearly, their overall sense of sight, a combination of their ability to see at night and also to detect infrared radiation, gives them an important hunting advantage. Another advantage critical to their hunting success is their highly developed sense of smell, which allows them to detect fresh rodent scent and thereby place themselves in areas where they are most likely to encounter a rodent in close proximity, such as alongside rodent trails and at the base of oak trees. Not only does their sense of smell allow them to get close to and kill their prey, it also makes it possible for them to follow up the animals they envenomate.

How well developed is a timber rattlesnake's sense of smell? I am under the impression from my own personal experience with these snakes in the field that their sense of smell is exceptional. On several of the field trips I have taken into timber rattler summer habitats with W. S. Brown, we have heard snakes start rattling when we were well over twenty feet away and approaching their positions with a fair amount of stealth from downwind. Did these snakes see us first or smell us? Given the position of the rocks under which we eventually located them, it seemed impossible that they could have seen us, or picked up any ground vibrations we were making at the point when they started to rattle.

The scientific community has known for half a century or longer how rattlesnakes derive their sense of smell. They get it through both their nostrils and their tongues, which are able to pick up sensation-affecting particles in the air and transfer the sensation of smell to their brains via the Jacobson's organs in the roofs of their mouths.[14] What has not been known by biologists and herpetologists is how highly developed this sense is in these snakes. Finally, some very interesting data are beginning to emerge from Dr. David Chiszar's rattlesnake research at the University of Colorado.[15]

Dr. Chiszar conducts an experiment in which a mouse is immediately removed from a cage after being bitten and envenomated by a rattlesnake. It is then dragged along the floor, out of the rattlesnake's sight, establishing a scent trail. At the same time the scent of a litter mate or mates of the mouse is dragged along the floor to establish conflicting trails. Afterward the snake is let out of its cage, whereupon it immediately begins to flick its tongue to pick up the scent trail of the mouse it envenomated. In 85% of these experiments the snake not only locates the trail through what is referred to as "strike-induced chemosensory search," but is able to stay on it no matter how many times it might be crisscrossed by the other mouse trails.

Thus we infer that the snake has an excellent sense of smell. Only two things differentiate the envenomated mouse trail from the other trails: the envenomated mouse's scent compared to the scents of the other mouse or mice, and the scent of the venom itself. The majority of rattlesnakes in this experiment are able to detect these subtleties, and this is what keeps them on course. It could be unproductive to follow the paths of unbitten rodents that either intersect or run adjacent to the paths of rodents they have bitten, and nature has equipped all rattlesnakes, not just timber rattlesnakes, with a strong enough sense of smell to prevent them from doing so.

Lumping all of a timber rattlesnake's physical traits together, it's easy to understand why they are such highly evolved predators. Even on the darkest of nights, a small rodent's life is in serious jeopardy if it gets too close to a timber rattlesnake in its hunting mode. Amazingly enough, on top of everything else this reptile has going for it in terms of hunting advantages, there is one additional advantage, not physical in nature, that seldom if ever gets discussed. I'll call it staying power. If timber rattlers were felines or humans, I'd call it patience. It's their ability to wait in one spot with minimal movement for many hours for the opportunity to make a kill. When it comes to waiting-without-moving skills, it's hard to imagine any creature on earth that could outperform a timber rattlesnake. When Art Moore, the legendary New York State timber rattlesnake bounty hunter, was working for the Department of Conservation in pest control throughout the state in the 1960s, it was not uncommon for him to pass through the same area of woods day after day. He told me, in one of my early interviews with him in the summer of 2002, that he frequently saw timber rattlers at the base of oak trees, and often one would be at the base of the same tree for several days before moving on.[16] Whether they scored or not was uncertain, but their long vigils were a real testament to their staying power. I think it is safe to assume that a lot of those snakes scored. They wouldn't have been at the bases of those trees in the first place if there wasn't fresh rodent scent promising rodent activity.[17]

I have emphasized that the timber rattlesnake's senses of sight and smell have given it a great advantage when it comes to hunting during the hours of darkness. Implicit in emphasizing this advantage is the fact that these snakes do at least some of their hunting at night. The fact is that timber rattlesnakes do a lot of their hunting at night, but not throughout their entire active season. It's more common during the months of summer than in the spring or fall. It all boils down to the fact that timber rattlers are sensitive to a high degree of ambient heat and sunshine at the same time. On sunny days, once the air surrounding them becomes uncomfortably warm, they tend to spend their time under overhanging rocks, or in shaded forests, or almost anywhere out of the sun. If one thinks of these snakes as sun-loving creatures, it may be difficult to think of them as night hunters and creatures of the night in general, but that's exactly what they are, even in the northern fringe of their range in the northeast, when the air gets hot.

On a spring or fall day when the air temperature is below 75°, the ground is cool, and the sun is shining, one is likely to find timber rattlesnakes basking in the sun.[18] They are attempting to bring their body temperatures up toward their optimum range, which is probably somewhere between 80° to 85° Fahrenheit.[19] During the warmest periods of summer in Vermont and northeastern New York, favorable basking conditions might only occur in the first few hours of daylight. This can also be the case in other areas within the timber rattlesnake's range, such as W. H. Martin's high-elevation study areas on the Virginia/West Virginia border.

I have accompanied W. S. Brown on numerous afternoon field trips to his study area in Warren County, New York, during July and August. We always locate timber rattlers at this time of day, but we never find them basking in the open. Even in areas where they're known to bask, they always seem to be in the shade. That's because, by the time we discover these snakes, the combination of ground temperature, air temperature, and sunshine is already beyond their comfort level. I am unaware of any timber rattlesnake studies that have estimated the point at which timber rattlesnakes transition from comfortable to uncomfortable on hot summer days. An educated guess is that it might be somewhere in the high 70° range with the sun shining, and possibly somewhere in high 80° range without sunshine.[20]

People often use the words "strike" and "bite" interchangeably. A good way to think of a strike is to envision a snake as it throws itself forward at an intended target. In a defensive strike, timber rattlesnakes are capable of hitting a target half their body length away. Their predatory strikes are typically much shorter. As previously mentioned, these snakes tend to hunt next to rodent trails and in other areas with fresh rodent scent. Therefore, when a timber rattler strikes at prey, its strike is often as short as six to eight inches.

Timber rattlesnakes are able to deliver their venom into an intended target in two ways. It all depends on the size of the target. If the target is relatively small, such as a human hand or foot, or a small rodent, they propel themselves forward with their mouths open, and bite down after making contact. If it's a much larger target such as the leg, body, or head of a human being, they open their mouths as wide as possible, rotate their fangs outward, propel themselves forward, and inject their venom after making contact. In this type of delivery, they don't need to bite down at all. It's conceivable that a timber rattlesnake could envenomate a mammal of any size if it were within striking range.

THE HUNTER AND THE HUNTED

Among the many small rodents that timber rattlesnakes feed on, it appears that the white-footed mouse and the eastern chipmunk are their two primary prey species in Rutland County, Vermont, and in Warren, Washington, and Essex counties across Lake Champlain in New York. We know this through W. S. Brown's field study, which commenced in 1978 and is still underway at the time of this writing, in the eastern foothills of the Adirondacks in New York.[21] If this is in fact the case, it's probably due to the fact that these two rodents are more available than other species of rodents in the northern Appalachian oak-hickory forests that predominate in his study area as well as in western Rutland County. It's important to realize, though, that a timber rattlesnake is a feeding opportunist that will prey on any one of a wide variety of rodents if the opportunity arises. By examining the stomach contents of over eleven hundred preserved specimens from over ten museums throughout the east and midwest, Rulon Clark discovered the remains of ten different mammals, including shrews, short-tailed shrews, house mice, golden mice, New World mice, voles, red-backed voles, cotton rats, eastern chipmunks, and gray squirrels. He also found the remains of cottontail rabbits in some of the adult snakes he examined.[22]

In one of my interviews with Art Moore in 2002, I asked him if timber rattlesnakes hunt other mammals in addition to mice and chipmunks. Moore, a prolific hunter and killer of timber rattlesnakes, was also a serious student of the snakes. He began his answer by claiming that he once killed a noticeably lumpy timber rattler in a hayfield in Washington County, New York. After cutting it open to see what it had eaten, he was surprised to find not one but seven star-nosed moles inside. He went on to tell me that, in addition to moles, chipmunks, and countless mice, he had seen a remarkable variety of other young animals inside snakes he killed, including juvenile rabbits, gray squirrels, muskrats, and groundhogs. He also discovered the remains of adult red squirrels and flying squirrels on occasion. I mentioned to him that I had only ever seen one flying squirrel in my life. His answer came quickly and has stuck with me, "You aren't gonna see too many 'cause they're nocturnal." As avid an outdoorsman as Moore was (he passed away in the summer of 2003), I have no reason to believe that he didn't find the various remains he claimed that he did.

It's a fair assumption that timber rattlesnakes get enough to eat in the wild because of their hunting prowess, but how much is enough? There's probably nobody around who can definitively answer this question. W. S. Brown estimates that timber rattlers eat between six and twenty times a year in the wild. He points out that a number of factors can affect how many times they need to eat a year in order to survive. Larger animals, for instance, will give them more sustenance and stored energy than smaller ones. It could easily take several white-footed mice to equal the amount of energy they would derive from eating a young gray squirrel or rabbit.

It's also fair to assume that timber rattlesnakes drink enough water in the wild to survive, but how often and how much do they need to drink? Brown claims that, as long as they're eating well, they're getting enough water through the animals they ingest and that they can go without actual water for weeks on end, perhaps even to the edge of dehydration, and still survive. To illustrate this point, he refers to a coiled timber rattlesnake he once watched on a ridge during a brief but intense summer thunderstorm. It hadn't rained in six weeks, and the mountain was bone dry. As the heavy rain tapered down, the snake began to drink water from between its coils. This account may not address the issue of how often and how much these snakes need to drink in order to survive (surely it's as critical to them as it is to any other living creature), but it tells us something about how critical it is, or isn't, for them to be near water. Even in a severe drought, they don't have to be near water or seek it out. One way or another, water is likely to come to them.

It's clear that timber rattlesnakes are hard-wired for survival, and that they play an important role in food chains of the forests they inhabit. I believe that Art Moore summed up their role as well as anyone when he told me in one of our early discussions that timber rattlesnakes specialize in rodent control. There's no question that they kill a lot of rodents. Conversely, a lot of animals, including deer, horses, cattle, domestic dogs, and human beings, are capable of killing timber rattlesnakes, but my focus here is on animals that kill them as a source of food.[23] A wide variety of animals are known to kill and eat these snakes if they get an opportunity to do so. They attack from the ground as well as from the air. Some walk, some fly, and at least one slithers.

One of these predators is the black racer. Randy Stechert has been in the field with both snakes since the 1960s and knows them well. He confirms that black racers are predators of timber rattlers, and gives an account in which his friend, a naturalist named Roy Pinney, witnessed an adult black racer attack and swallow several neonate (newborn) timber rattlers that were in close proximity to each other on a den in southern New York one fall.[24] From Stechert's point of view, the most interesting detail in the account is the way in which the racer reacted after coming upon the young, twelve-inch-long rattlesnakes. Apparently sensing that it might be able to kill and eat only one of the little snakes while the others took cover under rocks, the racer grabbed one of them in its mouth, chewed on it, dropped it, then moved quickly forward to another one. It repeated exactly what it had done to the first one. Before the little reptiles were able to safely scatter and hide from their attacker, the racer had grabbed, bitten down on, and dropped about four or five of them, all of which were injured to the point where they couldn't escape to safety. When there were no more neonates to attack, the racer methodically went back to all of the snakes it had disabled and swallowed them headfirst.

Stechert feels that black racers are much more likely to kill neonates than older timber rattlesnakes, and that the major reason for this tendency has to do with the growth rate of timber rattlers. He has found over the years that timber rattlesnakes are about twelve inches long when they're born. As one-year-olds, they are apt to be somewhere between eighteen and twenty inches in length. By two, they're usually twenty-one to twenty-four inches long. A year later, as three-year-olds, they typically run between twenty-six and thirty inches, and once they become four years of age, they're often in the thirty-four to thirty-six inch range. These growth rates may not be the norm for these snakes throughout their entire range, but they illustrate a point. Where a four- to five-foot-long, slender, adult black racer weighing approximately a pound might be relatively comfortable subduing and eating a baby timber rattler, it might hesitate to deal with this species as it ages, lengthens, and puts on weight. A newborn timber rattlesnake weighs about twenty grams, an ounce or less, but a well-fed, thirty-inch-long three-year-old could be in the neighborhood of half a pound, and a three-foot-long four-year-old could weigh a pound or more.[25]

I have mentioned that domestic dogs are capable of killing timber rattlesnakes. Coyotes and foxes are too. All three canines tend to kill them the same way. They wait until the snakes are uncoiled and moving away from them, then move in quickly, grab them behind the head, and shake them to death. Where all three canines kill rattlesnakes by instinct, there is evidence that coyotes and foxes kill them as a source of food. Laurence Klauber refers to some cases on the west coast where rattlesnake remains were found in the stomachs of coyotes, and also at den sites.[26] In one case there were rattlesnake heads at a den site, leading a U.S. Fish and Wildlife Service employee to surmise that the parents had brought the snakes to the den for their pups to eat. Although the rattlesnakes involved in these cases were not timber rattlers, the cases tell us something about coyote behavior. It seems very likely that a coyote wouldn't make a distinction between a Great Basin rattlesnake (*Crotalus oreganus*) in Idaho and a timber rattlesnake (*Crotalus horridus*) in Rutland County, Vermont, or anywhere else.

Although I am unaware of any written references to coyotes preying on timber rattlesnakes, W. S. Brown has gathered an interesting piece of circumstantial evidence pointing to the fact that these canines will eat timber rattlers, at least on occasion. At one of the spots he visits annually in his study area, he found a mortally wounded adult timber rattler coiled and sunning itself on an open grassy knoll in front of a large rock that the snakes use every summer for shelter and basking purposes. The snake, an adult male black morph, had obviously been attacked by a predator. It had a gaping hole in its mid-body region, exposing its heart, lungs and stomach. As Brown briefly studied the snake, he noticed that something had dug out the grass and soil from the entire perimeter of its rock sometime prior to his arrival. Although there were no obvious tracks in the fresh soil, he imagined the scenario that may well have played out there: The rattlesnake had been attacked by a coyote, a species of canine that abounds in his study area. Even though wounded severely, it had managed to get away from its predator and took refuge under its rock. At that point the coyote had made quite an effort to expose the wounded snake and drag it back out, but was unsuccessful, and wandered off. Brown told me that he went back the following day to check on the wounded snake. Amazingly, it was out basking normally, even though it wasn't in a normal condition by any means.[27]

Randy Stechert and W. H. Martin recount having a somewhat similar experience on a field trip they took together in Maryland. They also came upon a large rock with obvious signs of pawing and digging all around it. Their rock didn't feature a wounded, basking timber rattlesnake, however, but a litter of timber rattler neonates, with canine tracks obviously made by a fox. After studying the area for a few minutes, both men concluded that a fox had noticed or scented the young rattlesnakes, some or all of which had managed to escape under the rock. The fox then proceeded to paw and dig extensively around their rock in an attempt to expose and eat any of them that it could. Whether it had been successful or not was uncertain.[28]

A fair amount has been written about the red-tailed hawk as a predator of rattlesnakes. Their technique is to sink their talons into the head and neck areas of the snakes, thus rendering them defenseless, and then proceeding to tear them apart with their beaks and eat them. Very often these birds will fly away to their nests and other areas after swooping down and grabbing a rattlesnake. Some accounts tell of red-tailed hawks seen flying with rattlesnakes in their talons, sometimes quite sizable ones. There is one report of a man in Arizona shooting at a red-tailed hawk that was flying with a large rattlesnake in its talons. The shot caused the bird to drop the snake, which turned out to be four-feet-two-inches long and approximately two-and-a-half pounds.[29]

I am unaware of any written references to red-tailed hawks killing and/or eating timber rattlesnakes per se, but Randy Stechert has some strong circumstantial evidence suggesting that these birds are capable of preying on timber rattlesnakes. For a number of years he noticed what he thought to be the same female red-tailed hawk perched nearby a rocky area where female timber rattlesnakes gave birth in early September annually. He saw this bird so regularly during his field trips that he became accustomed to finding it during his visits, not only to the birthing area in early September but also to the den itself a few weeks later in late September and early October. One year he failed to find the female, but in her place were two young red-tailed hawks that he presumed might be the offspring of the bird he had seen so often. Whether any of these birds were related, Stechert will never know. It's not even certain, he admitted, that the adult female he had seen so frequently and often in almost the same settings, was in fact the same female year after year. What he has inferred from his observations of these hawks is that they were in the hunting mode whenever he came upon them, waiting to swoop down and grab a meal whenever the opportunity presented itself.[30]

Where many people might not find it unusual that red-tailed hawks, and other hawks as well, are known to kill and feed on rattlesnakes, there are some who might find it hard to imagine that the wild turkey is known to prey on these snakes. Dr. Daniel E. Keyler, a timber rattlesnake expert at the University of Minnesota, related an experience to me in which he witnessed something that probably very few humans have ever seen, a wild turkey gobbler subduing and eating a timber rattlesnake. In August 1997, Keyler was in some of his favorite timber rattlesnake habitat in Winona County, Minnesota, when he came upon a wild turkey gobbler only twenty-five yards away along the edge of a gravel road. The bird was preoccupied and hadn't noticed his approach. Keyler froze in his tracks and quickly discerned that the gobbler was holding down a twenty- to twenty-four-inch-long timber rattlesnake with both feet. The snake was striking repeatedly at the bird's long, armored legs and folded-in wings, but to no avail. It was about ten-thirty in the morning. The sun was shining brightly, and Keyler was in for one of the rarest nature sightings of his life. When the snake stopped striking, the gobbler reached down and grabbed it behind the head with its beak. While continuing to hold the snake down with its feet, it proceeded to pull upward, stretching the rattler's neck to the point where it appeared to rip open. This tugging went on for about two to three minutes. Finally, with the snake zapped of its energy, the big bird picked it up by its head and began to work it into its crop. After a minute or two of "juggling," as Keyler described it, the snake totally disappeared from view, and the gobbler began to move off.[31]

ANNUAL CYCLE AND HABITAT

Those reading about timber rattlesnakes for the first time may be surprised to learn that these snakes have a recognizable and purposeful pattern of movement in their lives, referred to by those who know and study them as their annual cycle. This cycle begins with a period of hibernation that is as long as seven months in the northern fringe of their range. Hibernation is followed by what's known as emergence, or the period of time in the spring when the snakes come up out of the ground and begin to congregate in numbers near the entrances to their dens. Once the weather warms up enough, they gradually begin to egress (migrate) away from their dens for summer foraging, mating, and birthing in the case of those females that are reproductive in a given year. In their migrations it's not uncommon for timber rattlers to reach points that are two to three miles away from their dens. Toward the end of summer, they begin to ingress (return) to their dens to start the cycle again.

As a rule, the dates of spring emergence in Vermont and northeastern New York are later than they are in many parts of the central Appalachians and points farther south. There are some exceptions to this in high elevation areas, as on the Virginia/West Virginia border. We know that elevation can have a major effect on shedding rates. It can also play a major role in the arrival of spring emergence. Springtime can be painfully slow to arrive in Rutland County, Vermont, and Warren, Washington, and Essex counties in New York, and the month of April can be extremely chilly and raw. By the end of April and the beginning of May, some 60° to 70° weather finally begins to settle in. This is when the timber rattlesnakes in this area first begin to emerge from their long period of hibernation. Being cold-blooded reptiles, they need temperatures in this range to become active.

Once they start emerging, their tendency is to bask on or near the surface of their dens on warm sunny days and go back at least partly into their dens at night, when the temperatures often drop into the thirties. Dropping temperatures during the day can also send them scurrying for cover. They tend to stay close to their dens in a pattern of basking and taking cover from the cold for a week or two, and in some cases longer. Eventually, days with temperatures in the sixties and seventies followed by nights in the forties and higher lock in, and the snakes begin to migrate out toward various destinations within their summer habitats.

In some cases their migrations can exceed three miles in length. I know of one confirmed report in Washington County, New York, in which a timber rattlesnake with a recognizable painted rattle traveled a little under four miles from its den. John McDonough, a licensed New York wildlife volunteer for the Department of Environmental Conservation (DEC), confirmed with me in a conversation on February 22, 2004, that a rattlesnake—whose rattle he had spray-painted and which could thus be traced to a particular den in Washington County, New York—was found nearly four miles away from its den one summer. At this time, W. S. Brown has the record confirmed migration distance for a timber rattlesnake. In the summer of 2004, he found a road-killed timber rattler, one of his marked animals, 4.5 miles from its den in Warren County. The fact that the snakes in both of these cases were males isn't surprising, as the male of this species is known to migrate the longest distances in its search for mates. Female timber rattlers don't typically go as far from their dens in the spring, and when it comes to mating, they don't have to locate males. Males find them.

It can be a little difficult imagining that such lowly creatures as timber rattlesnakes can migrate to points up to several miles away after emerging from their dens in the spring, only to turn around and return to their dens in the fall. This is no less remarkable to me than a tiny bird flying a thousand miles away in the fall, and returning to the same territory where it was born the following spring. If more people understood that timber rattlesnakes are migratory animals, I firmly believe that far fewer of these snakes would be needlessly killed every year. The vast majority of timber rattlers killed on private properties have come from dens up to a mile and more away and are merely attempting to pass or migrate through the area.

Timber rattlesnakes have adapted to a wide variety of habitats throughout their entire range. Probably the best way to study the overall habitat in which any population, or colony, of timber rattlers lives is to first locate its winter habitat or den.[32] Studies in which these snakes are tracked via various techniques including radiotelemetry (the process of tracking creatures that have been implanted with a tiny transmitter that sends out a signal that can be picked up by a receiver several hundred yards away) have shown that their entire lives are apt to be spent within a mile or two distance from their dens. Timber rattlesnake experts are able to home in on the snakes not only at their dens but also during the months when they're out of their dens and widely dispersed in areas as wide as two to four miles across. These men are not magicians. They simply understand the annual cycle and bodily needs of timber rattlers, and are able to fan out from dens and locate the snakes, even in areas with which they are totally unfamiliar.

The similarity of their denning locations is one of the most interesting aspects of the border rattlers of western Rutland County and the eastern foothills of the Adirondacks. Invariably, these locations are in what are referred to as talus slopes, or rockslides, which in some cases can be seen from a long distance away. Wherever they occur, talus slopes are apt to be pronounced features on the side of a mountain, and have a remarkably similar appearance. They are large, sloping piles of rubble below steep cliffs that usually face in a southerly direction. Talus slopes can be an acre or more in size and support almost no vegetation other than scattered small trees and saplings and minimal ground cover such as wild grape vines, Virginia creeper, and poison ivy. Invariably, they have their biggest boulders at their bottoms and their smaller rocks and stones at their tops. Timber rattlesnakes typically locate their dens in deep cracks along a line where the rubble and the cliffs above them meet. The cracks must be deep enough to allow the snakes to get below the frost line. This is the key to their winter survival; they need to maintain a body temperature in the upper thirties or higher while hibernating.[33] So the higher up a mountain their dens occur, or the farther north the snakes have established themselves, the deeper the cracks have to be to prevent them from freezing to death.

Raymond Lee Ditmars, a well-known herpetologist and reptile curator in the first half of the twentieth century, surmised that talus slopes were formed as the result of cataclysmic forces in the distant past.[34] When looking for geological forces capable of causing a large area of mountainside to break away from the main body of a mountain, one must consider the colossal power of glaciation. In one of my conversations with Bill Galick, arguably Vermont's top timber rattlesnake bounty hunter, I asked him what caused the massive talus slope on the mountain above his two-thousand-acre farm in Rutland County. His answer came quickly, "Ice."[35] W. H. Martin, visiting the den at the Galick farm, confirmed that ice is undoubtedly what created the talus slope in which the den is located, and proceeded to give his scenario of how things may have transpired there geologically.[36] Initially, southerly movement by any one of the numerous ice sheets that covered the northeast in the geological past could have scraped a large area near the top of the mountain down to bare rock. If this didn't happen, the tremendous run-off from the ice sheets as they melted and retreated northward could easily have caused landslides and the subsequent exposure of rock. To visualize how powerful the flow of water may have been, consider that the most recent ice sheet to have covered much of the northeast and midwest at the end of the Pleistocene epoch is believed to have been between a mile and two miles thick. Whether the mountain's bedrock was exposed by the advance or retreat of a glacier, once ice no longer blanketed the area, a talus slope began to develop. This developmental process goes on even now at the Galick farm as a cycle of water trapped in exposed cracks or faults near the top of the mountain, forming ice during the winter months, which expands to break away fragments and boulders, which, little by little gather to form a huge pile of rocks below steep cliffs.

Although talus dens below cliffs predominate in Vermont and northeastern New York, they are not unique to this area. They occur in other parts of the northeast such as in Randy Stechert's study areas in southern New York, and in the areas that W. H. Martin studies in Virginia and West Virginia. Martin claims that they occur as far south as North Carolina but that they are not common there. Several other types of dens occur throughout the timber rattlesnake's range in the east, some in vertical or horizontal fissures in ledges, others in rockslides called scree that have no cliffs above them. Another type is referred to as a fallen rock den.

Fissure dens are common in the mountains to the south of the area I frequently describe as the border rattler country of Vermont and northeastern New York, and occur all the way down to Alabama. In this type of den, an opening between ledge and earth is deep enough and large enough to protect a denning group of timber rattlesnakes from predators and cold weather. W. H. Martin's research has shown that fissure dens, as well as what he terms partially covered scree dens predominate in the mountains of Virginia. Depending on their size, both fissure dens and scree dens are capable of holding healthy populations of timber rattlers. Although a count of thirty to sixty snakes per den has been the norm in most of the dens he has studied over the years, Martin estimates that a small percentage of the dens he studies hold as many as a hundred and twenty to two hundred snakes.[37] In the case of fallen rock dens, they're probably able to hold only very small populations or small numbers from diffuse populations of snakes at best.

A totally different type of habitat is found in a vast area of southern New Jersey known as the Pine Barrens. Here the terrain is largely flat and not at all similar to the rocky, mountainous habitat that timber rattlesnakes typically prefer throughout the northeast and the Appalachian Mountains in general. Studies have shown that the Pine Barrens timber rattlers meet their overwintering needs by denning up in pockets where tree roots and water-logged land come together at the heads of streams within or near white cedar swamps.[38] A good term for these locations, in my opinion, is root-pocket dens. The fact that such a denning solution is used by the timber rattlers of this area shows us something revealing about this species' ability to adapt to its surroundings.

Once timber rattlesnakes leave the security of their dens in the spring, they tend to head to basking areas, or basking knolls, as they are known by herpetologists, as well as to foraging habitat. On their way to these areas they typically stop off briefly in what was described by Raymond Ditmars in 1931 as transient rocks and later called transient habitat by Randy Stechert, W. S. Brown, and others.[39] Brown's research in northeastern New York has revealed that transient habitats are usually within a couple of hundred yards of dens but in terrain that is lightly forested and visibly rocky. Usually located on outcrop knolls, they offer everything the snakes have left behind at their dens in terms of the ability to bask in the sun and to take cover. Although scattered with a variety of small trees, including red oak, shagbark hickory, hop hornbeam, white pine, staghorn sumac, and red cedar, these areas are open enough for the snakes to bask. Additionally, they invariably have rocks, which Brown refers to as shelter rocks, and/or rocky surfaces with crevices that the snakes can use for shelter. One of the main reasons that timber rattlesnakes use transient habitat in Brown's study area is for the protection it can give them against any late spring cold weather they might encounter after leaving their dens. By mid to late May the weather in the area has warmed up to the point where the snakes have become more active and moved on to their summer habitat.

As their name suggests, the bulk of these snakes end up in heavily timbered areas for the purpose of foraging and also mate searching, but some of the snakes leave the timber and seek out certain open knolls for the purpose of basking. Sometimes basking knolls are quite close to their dens and connected to or part of transient habitats, but they can also be a mile or more away from their dens. In either case, they are open rocky areas with relatively sparse ground cover and very few, if any, trees present. The major difference between transient and basking areas is trees. Otherwise, they can be almost identical in appearance.

In all of the field trips that W. S. Brown and I have taken to his study area in July and August, over 95% of the snakes we have located were under or next to large rocks on basking knolls, and in most of these cases the snakes we found were gravid (pregnant) females. They are typically in these areas because they need to maintain an optimum body temperature range in the high eighties to low nineties to bring their pregnancies to a successful completion. Females who are not pregnant, but whose eggs are developing, are likely to be found on basking knolls as well. Sometimes referred to as "yolking" females, these snakes also need to maintain an optimum internal temperature range to help develop their eggs. In addition to pregnant and yolking females, pre-shedding snakes of either sex (those preparing to molt or discard their entire outer skin covering) need to maintain an optimum internal temperature range and are therefore likely to be seen on basking knolls. And finally, a male timber rattler looking to mate with any receptive females that happen to be present may also make an appearance on a basking knoll.[40]

Where basking knolls are fairly easy to visualize and define, this is not the case with foraging habitat. If you are standing on or next to a timber rattlesnake den in any relatively undisturbed mountainous area, the foraging habitat of the snakes that hibernate in the den is essentially the entire mountain around you for two to three miles in every direction. Another way to put this is that the snakes from the den will forage in likely areas to find their prey in a circle up to five miles and more in diameter, and at the end of their active season (that part of the year in which they are not in hibernation) will return to their den to spend the winter.

Very few humans get the opportunity to see a timber rattlesnake kill and devour any of the many small animals the species is known to prey on in the wild. So we have to generalize a little when describing the specific foraging habitat of these snakes. It may be useful to think of it in the following way. It's any place, such as along the side of a fallen log, where a timber rattlesnake is seen lying motionlessly in its classic ambush posture (see pl. 2). It's any place where a timber rattlesnake is seen with obvious bulges in its stomach. It's probably safe to say that foraging areas are likely to encompass sections of forest in which oak trees either predominate or occur commonly. This is certainly the case in the oak-hickory forests of western Rutland County, and across Lake Champlain in Warren, Washington, and Essex counties as well. As mentioned previously, Art Moore saw timber rattlesnakes at the base of oak trees on many occasions. There's a reason for this. Where there are oak trees, there are often acorns. Acorns attract chipmunks and squirrels, which in turn attract timber rattlers. It is very likely that a timber rattler will detect a chipmunk's or squirrel's scent, follow it to the bottom of an oak tree, and wait for the unsuspecting animal to return.

Although these snakes conduct the bulk of their hunting in forests, they are certainly known to hunt in other areas such as freshly mowed hayfields that border hardwood forests. In Rutland County, below one of its three timber rattlesnake dens, there's a dairy farm on which one or two rattlesnakes are killed every couple of years during haying. According to the owner of the farm, the snakes come out of the hardwoods at the base of the mountain into an adjacent field almost as soon as he begins to mow it every summer.[41] It's likely they're going into the field to hunt mice, voles, and other small rodents. The lumpy snake with seven star-nosed moles inside that Art Moore killed in a hayfield in New York comes to mind. The field in which he discovered and killed the snake was being mowed at the time.

I have visited timber rattlesnake dens, basking knolls, and foraging habitat in both Vermont and upstate New York and am struck by the similarities, especially of the dens. They not only look alike, there's a certain feeling in the habitat around them that I haven't experienced elsewhere in the country. It's more than a feeling of remoteness. It's as if I have entered into a no-man's zone or a refuge where man seldom treads, and that I need to be on guard while there. It's not just the snakes that give me a premonition of danger. It's the steep and rocky terrain and the realization that help would be a long way away if I were to take a serious tumble and hurt myself. My hat goes off to the timber rattlers in these areas. Long ago they took these fortified hilltop positions, the rugged steepness of which has been a major factor in their survival.

There's no question that timber rattlesnakes migrate to various locations within their summer habitats as part of their annual cycle. In W. S. Brown's ongoing field study, in which he has captured, marked, released, and recaptured over two thousand of these snakes, he has learned a great deal about their migrations. Every time he recaptures one of his previously marked snakes, one of the many things he learns about it, after consulting the log book he always carries with him in the field, is whether it has been in that location before. Brown has long known that these animals are capable of returning to the same areas and even to specific shelter rocks within them annually. Other researchers have noted this tendency in timber rattlesnakes as well. But one of the things that sets Brown's study apart from others is the fact that his is the only one in which timber rattlesnakes are shown to migrate to the same islands in consecutive years.[42] A number of islands on one particular lake in the Adirondacks attract a few of these snakes every summer. One might wonder why they swim out to islands when there are miles of prime timber rattlesnake habitat in the rugged mountains surrounding the lake. The reality is that it's no more unusual for them to be on an island than anywhere else in a two-to-three-mile radius of their dens. The snakes are migratory and known to travel long distances for basking, mating, and foraging purposes. Since, like all snakes, timber rattlers are good swimmers, it is not unreasonable to assume that a relatively small number of them might swim, rather than crawl, to favored spots within their summer habitats each year.

It is one thing, however, for a timber rattlesnake to find its way back to the same summer habitat or anywhere else it's been before by land, and another thing for it to accomplish this same feat over water. On land the snakes use their strong sense of smell to pick up scent trails left by other snakes. Since scent is totally broken down by water, obviously other abilities come into play that enable timber rattlers to swim to given destinations. One of the most memorable timber rattlesnakes in W. S. Brown's study was a large male nicknamed "Champ," who liked to spend his summers on one of the popular camping islands on the lake referred to above. Brown remembers, and has documented in his notes, six different summers during which he was called by DEC campground rangers to come and rescue Champ on the island. In every case, he was more than glad to do so since it was the part he's expected to play in the "nuisance timber rattlesnake" program he established with the DEC in the early 1980s. By definition, a "nuisance rattlesnake" is any rattlesnake that shows up where humans don't want to see one.

The way the program works is quite simple. Whenever campers register to camp on any of the islands known for occasional timber rattler appearances, they are informed what to do if they discover a rattlesnake during their stay. They are supposed to inform one of the rangers. Once alerted, the rangers, many of whom have been trained in how to properly handle a timber rattler, capture the snake, place it in a designated drop-off barrel, and call Brown. Usually within a day or two he picks up the snake and releases it on the mountainside from which it came. Of the scores of rattlesnakes that Brown has removed from the various islands over the years, some have returned to the same island repeatedly the way Champ did.

Two interesting inferences about timber rattlesnakes can be drawn from Brown's experience with campsite snakes. One suggests a strong likelihood that these snakes are able to imprint on, or develop a memory for, places they have been in the past. It so happens that in several nuisance captures by the DEC campground rangers, the snakes were close to the same picnic bench in the same campsite where they had been captured in previous summers. Brown does not think this was sheer coincidence. For example, if Champ was initially drawn to one particular bench (there are several others on the island) because of fresh mouse or chipmunk scent underneath it, his hunting there was successful enough to entice him to return to that same location in subsequent years. Since Champ is one of only a few rattlesnakes Brown is aware of that was ever captured on the small island, it seems fairly unlikely that he followed the scent trails of other rattlesnakes to the same campsite once he arrived on the island, and the inference that can be drawn about his gravitation to the site is that he was able to remember it from one year to the next.

The second inference one might make from Champ's repeated returns to his island of choice is that timber rattlesnakes have something akin to built-in compasses, in addition to a memory for places they have been to before, as part of their overall navigational ability. Even though there are a number of islands of similar appearance in the immediate area of Champ's island that he could have gone to, it's interesting that he was never captured on any of them. He clearly seemed to want "his" island. He also seemed to have the compass skills that made it possible for him to get there. One problem with this inference, however, is that Champ's island is not far from the shore of the lake, and he may have always been able to see and recognize it before swimming out to it.

If we can't infer from Brown's history with campsite nuisance timber rattlesnakes that somehow they know what compass direction to take to get to places they have been to in the past, we can, in my opinion, make this inference from the snakes that get out to two miles and farther away from their dens in the summer and yet manage to return to them in the fall. Without an inner compass, how else could they accomplish this? It's seems highly unlikely that they would try, or be able, to return to their dens by following the same route they took away from them during the course of the summer.

HOW DANGEROUS, HOW AGGRESSIVE?

Time after time I have observed that people who are unfamiliar with the biology of timber rattlesnakes seem to have the impression that these snakes are extremely dangerous and aggressive. This is a major misconception. They are wrong on both counts. Timber rattlesnake bites without some form of human provocation are extremely rare, and bites resulting in death are even rarer. During the course of the twentieth century there was only one documented death stemming from a timber rattlesnake bite in the entire state of New York. The victim, Charles Snyder, was the head keeper in the reptile house at the Bronx Zoo under the curatorship of Raymond Ditmars. His fatal bite, which is discussed in more depth in chapter 10, occurred in a valley in the southeastern part of the state in 1930.

To my knowledge no records exist that can definitively tell us how many people are bitten by timber rattlesnakes each year, or how many people have died from their bites in the United States over the last hundred or so years. Numerous facts and statistics, however, shed significant light on these questions. In the United States approximately seven to eight thousand people are bitten by venomous snakes annually. This includes bites from all species of rattlesnakes, copperheads, cottonmouths (*Agkistrodon piscivorus*), coral snakes (*Micrurus* sp.), and many species of exotic snakes in zoos as well as private collections. Out of all of these bites and subsequent envenomations (injections of venom into tissue), approximately nine to fifteen people die, six to ten of these from rattle-snake bites, and even fewer from timber rattlesnakes. Mortality data derived from poison center records, personal communications, media-type materials, and medical examiner reports have suggested to Dr. Keyler, whose specialty is venomous snakebite envenomation, that the death rate in the entire country from timber rattlesnake envenomations was as low as one person every five to ten years during the last three decades of the twentieth century.[43]

Who are the people most likely to be bitten by a timber rattlesnake? They tend to be the people who come into direct contact with these snakes and handle them for any of a number of reasons. This group includes researchers (both in the field and lab), those who have the snakes in private collections, snake-handling religious groups in the Appalachian Mountains and rural south, and others, but almost never outdoor enthusiasts such as hikers, hunters, and naturalists, most of whom will never see a timber rattlesnake in the wild in their lifetimes.

A number of authors have written extensively on the phenomenon of serpent-handling in the southern United States. As one author, Thomas Burton, has pointed out in his book *Serpent-Handling Believers*, the people who handle venomous snakes for religious purposes are not practitioners of a snake religion per se. Rather, they are Christian fundamentalists who believe that taking up serpents without harm is a sign of their faith (Mark 16:18) and allows them to advance through the several stages of the Pentecostal experience, which includes regeneration, sanctification, and baptism of the Holy Ghost. Burton describes the remarkable, unfaltering faith of snake handlers in general. Snakebite doesn't appear to deter these believers in the slightest. Many members of serpent-handling churches have atrophied hands and missing fingers stemming from bites. One of their famous ministers, Dewey Chafin, of Jolo, West Virginia, was bitten over one hundred times in his long career, and scores of others have died from their bites over the years.[44]

It is not my intention to digress into the psychology of serpent handling (that is, what attracts people to it and what they are able to derive from it), but rather to point out that seventy-four known deaths from snakebites over the last hundred or so years have occurred in churches following this religious practice. We'll never know exactly how many of these deaths were caused by timber rattlesnake bites, but since it was and still is uncommon for these churches to use snakes other than timber rattlers in their services, it is safe to say that the number of timber rattlesnake deaths in this country since 1900 would be considerably lower if religious serpent handling had never developed. The reason that most, if not all, of the timber rattlesnake deaths that have occurred in these churches were fatal is that the members who are bitten don't generally seek out medical assistance, but rely purely on their faith in God to counteract the toxic effects of their envenomations.

The incidence and experience of venomous snakebite appear to differ between males and females. In a recent analysis of four hundred and fifty venomous snakebite reports over a three-year period, the Arizona Poison and Drug Information Center was able to determine that males were bitten more often by venomous snakes than females were, the bites occurred more often on their hands and upper extremities, and the majority of males bitten were in the age range from twenty to thirty-five. What this seems to indicate is that young males take more risks when handling venomous snakes, and that they probably handle them more frequently than females do. In this analysis there was virtually no difference between the sexes in terms of accidental bites, presumably while in the field, in the lower extremities.[45] Although these cases included snakebites from all the species of North American venomous snakes and various species from other parts of the world, it's fair to assume that the male/female differences shown in the analysis are likely to apply in timber rattlesnake bite cases as well.

By emphasizing the fact that timber rattlesnake bites rarely prove to be fatal, I don't want the reader to assume that being bitten and envenomated by one of these snakes cannot be a serious medical situation. Many bites are extremely painful and necessitate an infusion of antivenom in a hospital to counteract the effects of the snake's venom. As serious as these bites can be, however, as I detail in chapter 3, they are far less life-threatening than many people realize. The people who die from these bites tend to be the rare individuals who develop a serious allergic reaction to the venom, or fail to seek out medical assistance. Everything considered, a person is more likely to die after being stung by a bee than bitten by a timber rattlesnake.

I'm not sure how so many people have developed the misconception that timber rattlesnakes are aggressive by nature. Before coming to know the behavior of these snakes, I assumed they were aggressive myself. The classic image of a coiled rattlesnake with its head reared back and ready to strike may have had something to do with it. In my initial interview with Bill Galick at his farm in Rutland County, I wanted to get his opinion on how aggressive he thought these snakes are. He claimed that he had never had a timber rattler strike at him in his entire career although he had stood within inches of them on numerous occasions. He attributed this to always wearing hip boots, which kept his body heat from being detected by the snakes. This may have had something to do with it, but I realized there must be more. That's when it first occurred to me that the species might not be aggressive at all.

In one of my first interviews with Art Moore I asked him for his thoughts on the subject. He told me that timber rattlers for the most part are passive snakes, but on occasion females, particularly before giving birth, can be feisty. Art apparently never encountered an aggressive male timber rattler in his entire career. I found this difficult to imagine since he had killed so many snakes. He pointed out that most of the thousands of timber rattlers he encountered as a bounty hunter never had a chance to exhibit their behavior. He killed them as soon as he saw them. It was during approximately a ten-year period of time in which he captured these snakes alive for sale to private collectors that he began to learn about their temperaments. He claimed that about 90% of them tried to elude him and that the small percentage of the remaining snakes that showed aggression toward him were invariably females.

When learning the biological facts about any species, no amount of conversation or reading can beat first-hand experience. It wasn't until I went on my first field trip with a bona-fide timber rattlesnake expert and was able to see my first timber rattlers in the wild that I realized how passive they can be. It was the tenth of October. W. S. Brown cautioned me when I arrived at his home in New York that the conditions were not ideal for finding timber rattlesnakes basking on their dens. Whereas it had been sunny and on the mild side in the late morning, by early afternoon clouds had moved in and the temperature was dropping rapidly.

"We should have gone a week earlier," he lamented. Despite the weather, which had turned raw enough to make the snakes seek cover, Brown was determined to introduce me to the excitement of being at a den when timber rattlers are present. I had previously been to dens with former bounty hunters at the height of summer and not seen any snakes at all. Somehow, with Brown I was confident that we would see at least one snake, and one was all I needed to make the trip successful in my mind. In one respect our timing was perfect. In northeastern New York and Vermont, most if not all timber rattlesnakes are back at their dens by early October.

After a brief conversation behind his house, we drove to the base of a rugged mountain range off in the distance, parked, and started what turned out to be a half-hour ascent with only two short breaks to catch our breath. It didn't take long to learn that, when it comes to walking up mountains to get to timber rattlesnake dens, Brown is totally fit for the task. It's fair to assume that any timber rattlesnake researcher in the Appalachian Mountains would have to be. Finally, some giant boulders began to appear on the side of the old logging road we were ascending, and I sensed that the den was nearby. Within another few minutes we were at the base of a large talus slope, and Brown didn't have to tell me that the den was located at the upper part of the slope at the base of the rugged cliff that towered above it. The realization that we were in timber rattler country hit me with a surge of adrenalin.

We made our way carefully up the steep rockslide and were within thirty feet of a section of cliff when Brown exclaimed, "There's one. Do you see it?" A long black morph timber rattler was moving slowly uphill only a few yards in front of us. Brown stepped forward quickly and gently picked the big snake up with his snake hook. I couldn't help noticing how relaxed it seemed while he was examining its rattle. It was draped over his hook with its head pointed away from us. All it seemed to be interested in was getting back down into the leaves and rocks. During the next half hour this scenario continued. We'd spot a snake. Brown would pick it up, examine it briefly, and release it. Then we'd move forward about fifteen yards and discover another snake. We saw ten snakes altogether including two neonates, and all of them were extremely passive. Our experience with those snakes that afternoon changed my perception of timber rattlesnake aggression completely. Even though timber rattlesnakes do tend to be passive in cool weather, it was really surprising to me to see just how relaxed and totally passive they where. I heard hardly a rattle in the forty minutes or so that we were on the den, and none of the snakes slipped back immediately into the rocks upon detecting us.

Brown's assessment of these snakes' nature is that they are essentially a passive snake that can be defensively aggressive when provoked. Unlike Art Moore, he has seen in his career plenty of males as well as females that were aggressive. In the field, he is as gentle as he can be with the hundreds of snakes he bags and processes but, as careful as he is not to hurt them in any way, he inevitably gets them a little riled up. It's an unavoidable part of his job. Many of these snakes take exception to being restrained by Brown as he steps carefully on them while the bulk of their bodies are within his collecting bag and about a foot or so of the posterior part of their bodies are outside of the bag.

This restraint can go on for ten minutes and longer while he does various things to them, such as clipping a number in the belly scales of snakes he hasn't previously captured, or palpating females to see what stage of development their eggs are in, or possibly drawing a few drops of blood from their tails for DNA analysis. If they take exception to being restrained, they can become aggressive. This same type of defensive behavior is probably what causes timber rattlesnakes occasionally to bite people in the serpent-handling churches of the south.

Ironically, timber rattlesnakes are often chosen by these churches because of their passive nature compared to other rattlesnakes, copperheads, and cottonmouths. Timber rattlers seem to have a limit, however, as to how long they can be gently handled or restrained, and they can change from passive to aggressive in an instant. They can also become defensively aggressive if molested, as any animal might. If you step on one accidentally, it may strike at you. On the other hand, if you step close to one, there's a good chance it will either do nothing at all or simply rattle.

At one point during my first field trip with Brown we both heard rattling and thought the sound was coming from about ten feet ahead. As Brown moved forward to locate the source of the rattling, I froze in my tracks, realizing that the buzzing sound was much closer to me. It was coming from a timber rattlesnake within a foot and a half of my boot, tucked into a little recess right where the top of the talus slope met an imposing section of cliff. All I could see of the snake initially was about eight or so inches of its golden-colored side. The large male yellow morph, which Brown captured and briefly examined, could easily have nailed me but had only rattled and, by so doing, had taught me one of my first important lessons about the nature of these snakes.

MATING AND LIFE CYCLE

As we accumulate more and more timber rattlesnake knowledge, we learn that the males of the species court females during the mating season. Courtship is by no means limited to mammals and birds. Reptiles demonstrate this form of behavior as well, although it is not commonly observed by human beings. W. S. Brown, however, has witnessed timber rattlesnake courtship on more than one occasion and reports that it's not unlike male courtship in other species, our own included. It's basically an attempt by a male to impress, or woo, a female prior to mating. The way male timber rattlers try to impress and, presumably, physically stimulate prospective mates is by positioning themselves beside them and rubbing the length of their bodies with a series of jerky chin movements. Brown witnessed this behavior during a field trip near the peak of the mating season on August 10, 2002, in a shaded area of woods not far away from a basking knoll in his study area in New York. Apparently he let nature take its course; then, upon examining the snakes, he found them to be individuals he had permanently marked and released earlier. He noted right away that their rattles were both painted the same color, and realized that they were from the same den, over two miles away from their current location.[46] It's certainly not uncommon for snakes from a den two miles away, or even from different dens in many cases, to end up on or near given basking knolls in the summer. Randy Stechert tells of a radio tracking study by Dr. Edwin McGowan of the DEC in which he tracked an adult female traveling with a large adult male. Knowing whether pairings such as these end up mating would cast some interesting light on timber rattlesnake courtship in general.

Not only do male timber rattlesnakes court females during their mating season, they also are known to combat other males over females. A naturalist named Jed Merrow was fortunate to observe timber rattlesnakes in combat in northeastern New York. During the excitement of viewing the confrontation between the two combatants, he took thorough field notes in order to be able to write up accurately his observations of the event at a later time. Apparently, Jed was observing two males and a female all in close proximity when combat erupted between the males, one of which was larger and more dominant than the other. It began with the combatants raising their forebodies up vertically and partially entwining each other. Then an obvious struggle ensued in which both snakes tried to overpower and throw each other down. The bigger snake clearly won the encounter and chased the smaller snake away, but the smaller snake was persistent and returned to the area. His opponent saw him sneaking in and chased him off again. The mating drive must have been strong in the smaller rattler because he came back for more, only to find the more dominant male now copulating with the female. True to form and totally undeterred, the smaller snake went over to the female and tried to woo her with chin rubbing. Instead of trying to chase him off again, the bigger male ignored his rival's presence completely, and the female ignored him too.[47]

Timber rattlesnakes seem able to maintain a veil of secrecy when it comes to their various mating activities. During his long career as a timber rattlesnake researcher, W. S. Brown has seen obvious courtship behavior only twice and copulation once, and he has never witnessed combat behavior such as Jed Merrow observed. Herpetologists commonly refer to these fights as combat dances. Brown told me he would love someday to have the opportunity to see and photograph a combat dance, but he realizes that the odds of this happening are slim at best. It's not going out on a limb to think of a combat dance as a reptilian wrestling match to which human beings aren't invited.

Where sexually mature male timber rattlesnakes mate or try to mate every year, this is not the case with females in the northern fringe of their range in the northeast. W. S. Brown has discovered in his twenty-five-year-long field study in the foothills of the Adirondacks in Warren County, New York, that approximately 60% of the females he has captured, marked, released, and recaptured, are on three-year reproductive cycles while most of the rest are on four-year cycles. A small percentage, about 10%, reproduce every five years, and a very small percentage, about 3%, are on six-year cycles.[48] There's a reason that females in areas like the eastern foothills of the Adirondacks, western Rutland County, Vermont, and the higher peaks of the central Appalachian Mountains aren't mating and reproducing more often. It has to do with the energy it takes to produce their eggs. Because of their long periods of hibernation, they have a relatively short active season in which to forage and store up critical energy as body fat. Consequently, their eggs take three and more years to develop.

In many parts of their range in the deep south, where the period of hibernation is relatively short, female timber rattlesnakes reproduce every two years. With a much longer active season, they can eat more prey, store more energy as body fat, and are able to develop their eggs more quickly. There is definitely a strong correlation between female egg development, reproductive cycles, and geographical locations. A recent study, for instance, has revealed that the majority of female lance-headed rattlesnakes (*Crotalus polystictus*) in Mexico are on one-year reproductive cycles.[49] This is attributable to the fact that the active season of this species is so long. Where hibernation can have a profound reproductive effect on timber rattlesnakes in the northern fringes of their range, it apparently has very little or no effect on the Mexican lance-headed rattlesnake. Hibernation slows rattlesnake egg development in two ways. First, it prevents females from eating and storing up critical energy. Second, while the snakes are in hibernation in a greatly reduced metabolic state, the development of any eggs within them is interrupted or put on hold until the following active season. A lot of time in which eggs would otherwise be developing can be lost each year because of hibernation.

As mentioned previously, so-called "yolking" females, or females who are not gravid but whose eggs are developing, are among the various timber rattle-snakes that one is likely to find on a basking knoll at the height of summer. In any given year, it is this group of females who become receptive to mating, their eggs having reached a critical phase of development. Unlike their male counterparts who "run" all over the landscape in hopes of finding a receptive mate, yolking females are pretty much able to stay in one spot and have mates seek them out. It's common for yolking females to shed during the mating sea-son. As they begin to shed their skins, these females emit a powerful phero-mone, or chemical scent, to which males are strongly attracted.[50] The female that Merrow witnessed near the combat dance was just beginning to shed her old skin at the beginning of the fight. The whole process took about ten min-utes, and immediately afterward she began to copulate with the dominant male.

A question that sometimes occurs to those involved in timber rattlesnake research is whether there is some way in which males are attracted to yolking females prior to their shedding. Males are certainly known to appear on bask-ing knolls in July and August. Is this because they know through experience that receptive females are likely to be in these areas? Or do yolking females leave a powerful scent trail on their way to basking knolls that is not only stimu-lating to males but capable of lingering for weeks afterward? W. S. Brown be-lieves the latter is very likely the case. If we accept the scent-trail theory, it ex-plains why the two males in the combat dance that Merrow witnessed were in close proximity to a yolking female prior to her shedding and release of her mating pheromones. They weren't just hoping that she would become recep-tive; they knew that she would.

One of the most interesting reproductive facts about timber rattlesnakes in the northern fringes of their range is that after yolking females mate they don't become gravid, or pregnant, right away. This is because their eggs are not yet fully developed. It is not usually until early summer of the following year, at least in Vermont and northeastern New York, that their eggs reach full development. This is attributable to the fact that these females have come out of hibernation, had a meal or two, and built up enough energy as body fat to make ovulation possible. By this time, their eggs, now fully developed, are able to be fertilized by the sperm they have stored since the previous August. This remarkable ability nature has given them is often referred to as delayed fertil-ization.

After successful fertilization of their eggs, female timber rattlesnakes ges-tate for about three months before giving live birth. In the mountainous border rattler counties of Vermont and New York, this usually takes place in the first half of September, and the normal number of neonates in a litter is usually somewhere between eight and ten. The little snakes stay with their mothers for about ten days to two weeks, shed their skins, then are on their own. It's a tough world and a short period of infancy for the young border rattlesnakes up on the basking knolls and talus slopes of Rutland County, Vermont, and Warren, Washington, and Essex counties in New York. In October, some of the neonates, despite following the scent trails of adults, never make it to the sanctuary of a den, and are preyed on by such predators as red-tailed hawks, coyotes, and black racers.[51]

How long are timber rattlesnakes capable of living? W. S. Brown and W. H. Martin have been able to make remarkably similar longevity projections for this species, based on Brown's long-term field studies in northeastern New York and on Martin's studies on the Allegheny Front in West Virginia. Brown, having made a statistical probability study of field data, estimates that neonates have a 65% rate of survivorship in their first year of life and a 90% rate thereafter.[52] Martin's data indicate a 68% rate of survivorship in the first year of life and a 92% rate for adult females once they reach eight years of age and older.[53] These may seem like high rates of survival for this species but, upon plotting survivorship curves based on these percentages, one can see that they actually aren't. Assuming a 65% rate of survivorship in the first year of life, 6.5 snakes out of a litter of ten neonates will survive until the age of one. With a 90% rate of survivorship afterward, 4.6 snakes will survive until the age of five, but only 2.27 snakes will survive to the age of ten. Interestingly, the odds that one of these two snakes will be a female are fairly high, a fact, as I explain shortly, that is vital to the perpetuation of the species. Brown and Martin have known for some time in their research that the female-to-male ratio in any litter generally runs about 50%.

With only two neonates per litter surviving to the age of ten, one might wonder if this species is capable of significant longevity. The answer is that these snakes are indeed capable of living a long time, if they first can manage to survive to the age of ten. If one were to use Brown's data to chart a survivorship curve for a hypothetical litter of one thousand neonates, the following facts would emerge: 650 of the snakes would make it to the age of one. Out of these, 87.8 would survive to the age of twenty, and 30.6 would survive to the age of thirty. Twenty- to thirty-year-old snakes are certainly not unknown in the wild. Brown is beginning to recapture a few snakes every year that he knows to be over twenty years of age, and a few since the year 2000 that he knows are over thirty years old. There is no guesswork involved. These are invariably individuals that he has captured, marked, recorded, and released in the past, in many cases over twenty years ago.

Any time Brown recaptures a timber rattlesnake that he has captured many years in the past, it seems to mean a great deal to him not just in terms of his study, but also on a personal level. In the summer of 2004 he recaptured a female yellow morph that he had captured on four previous occasions over a long span of years beginning in 1982. If Brown was correct in assuming that this snake was nine years old in 1982, it was thirty-one years old in 2004. He recaptured a second old snake in 2004, a black morph male numbered 446, that he had previously captured on four different occasions dating back to 1981. If he was accurate in his assumption that this snake was ten years of age in 1981, it was thirty-three years old in 2004, making it the oldest timber rattlesnake he has ever documented in his study.

For several reasons, the timber rattlesnakes on the Vermont/New York border have not rebounded well since the bounty on them was discontinued in the early 1970s. I have mentioned the fact that they have fairly small litters, a fairly low rate of first-year survival, and that the bulk of the females in this area are on three- and four-year reproductive cycles. What I haven't mentioned is the all-important fact that the average age at which most of the females in this area reach their sexual maturity is between nine and ten years, and that they will have as few as one to two, and perhaps as many as three to four, litters in their lifetimes, depending on how long they are able to survive. In other parts of the country, female timber rattlesnakes reach their sexual maturity at a much earlier age. Studies in Kansas, for instance, have shown that they reach their sexual maturity as early as the age of four.[54]

If the majority of the female timber rattlesnakes in Vermont and northeastern New York were able to have their first litters at a much earlier age, the species would undoubtedly be rebounding in the area much better than it currently is. As it is, if at least one female out of a litter of ten didn't survive until the age of ten in this area, the species would be in serious trouble. It simply couldn't depend on the relatively small proportion of females that reach their sexual maturity before the age of ten to carry the burden of maintaining the species at its current level.

The populations at most of the dens in the four-county area I discuss throughout this book are now apparently stable, but only a few, primarily on state lands in New York, have shown any improvement.[55] This change is difficult to document because of the fact that it has taken place so slowly over a period of many years. In retrospect, when Vermont and New York dropped their bounties on timber rattlers in the 1970s and protected the snakes legally in the 1980s, those were critical turning points for the species in both states. As a result of bounty hunting and various other encroachments by man, which I detail in following chapters, timber rattlesnakes had been severely depleted, and in many locations driven to extinction. Fortunately for the snakes, enough caring and enlightened people in those days realized that the species was in trouble in Rutland County, Vermont, and in much of its range in New York, and decided to take action to do what was right for nature, even if this effort benefited such a loathed and misunderstood creature as a rattlesnake.

Among many misconceptions about timber rattlesnakes, none is more common than that about their length. Since there is very little about this subject in the literature, I made a point, when interviewing some of the old timber rattlesnake bounty hunters who had hunted in Vermont and New York, of asking them about the longest timber rattlers they'd ever killed. Lester Reed claimed that the longest snake he'd ever killed in Vermont was sixty-three inches, and had a thirteen-segment rattle. He was hunting with his daughter, Carol, the day they shot it near the end of a hunt back in the 1960s. They knew they had killed the biggest snake of their careers and were back home an hour later, eager to get an accurate measurement of their big specimen. Lester held it at arm's length with its rattle just touching their patio, while Carol took a tape measure and measured it as carefully as she could. It was obvious to both of them that the snake was special. There could have been a small margin of error in her measurement, but it's unlikely that she was off by more than two or three inches.

In one of my visits with bounty hunter Bill Galick at his farm in Rutland County, I asked him if he'd ever killed any exceptionally long timber rattlesnakes in his career. Without hesitation he told me that his brother, Ed, shot one that was over six and possibly closer to seven feet long when they were hunting together shortly after World War II at the den on their farm. They knew that the big rattler, which Galick described as having a black pattern on a yellow background, was special, but they never thought to measure it. Nor did they photograph it. I called Ed Galick in Washington County, New York, for his opinion about the snake. He said it was by far the biggest snake that he or Bill ever killed and the only one they ever took that was so golden yellow.[1] It really stood out in his mind. I asked him how long it was and he said, "I don't know, but it was a big one." I asked him if he could say without any doubt that it was over six feet long and he said he definitely could. When I explained to him that it's a shame they didn't measure the snake, or save its skin, or take some photographs of it, he said, "We didn't know any better."

According to Lester Reed and Bill Galick, therefore, there were five-foot-plus timber rattlers in Vermont approximately half a century ago. I later asked Art Moore, who was New York State's most prolific timber rattlesnake bounty hunter, having unofficially killed in the neighborhood of fifteen thousand timber rattlers in his career, if he had ever killed or seen a timber rattler over five feet long. Since he had bounty hunted for over twenty years in the eastern foothills of New York's Adirondack Mountains—not far from Rutland County, Vermont—it seemed fair to assume that any facts he told me about the rattlesnakes in his area would pretty much apply to Vermont's rattlesnakes in Rutland County as well. Moore claimed that there were timber rattlers over five feet in length in Warren and Washington counties in the middle part of the last century and that his best snake was seventy inches long. Dr. E. M. Riley, the Vertebrate Zoology Curator of the New York State Museum in Albany, had offered him a standing reward of five thousand dollars back in the early 1960s for a timber rattlesnake over six feet long. It apparently didn't matter to Dr. Riley whether the snake was dead or alive. He just wanted to see a timber rattler that long, preferably from New York State. Moore tried for the rest of his career to capture such a snake, but never succeeded. I asked him if he live-captured or shot his seventy-inch snake, and it turned out he shot it. He saw its head poking out from a hole, fired before he ever saw its body, then tugged it out of the rocks.[2]

After talking with Lester Reed, the Galick brothers, and Art Moore, I was somewhat surprised to learn from W. S. Brown that the largest snake he has ever captured in the wild was a fifty-six inch male that weighed almost four pounds.[3] Brown measured it in a squeeze-box, a process that doesn't harm the reptiles at all and produces a close to perfect measurement.[4] How remarkable was that snake? In his twenty-five-plus-year career in the field in northeastern New York, Brown has never captured another timber rattler that was over four and a half feet in length. Nor has he ever seen another one that appeared to be that long.

Randy Stechert is probably more familiar with the timber rattlesnakes of New York than anyone else in the entire state. As one of the DEC's top field biologists, he is in the field gathering data on New York's timber rattlers almost daily from the earliest days they start appearing on their dens in the spring until the final days in which they can still be seen on their dens in the fall prior to going into hibernation. Besides having an impressive overall knowledge of New York's timber rattlesnakes, Stechert keeps abreast of timber rattlesnake data from other states in the northeast as well. In all of his over forty years in the field as both an amateur and a professional, he has only captured one timber rattlesnake that was five feet long. This occurred in one of two counties in southern New York that have the distinction of harboring the largest timber rattlers not only in New York State but in the entire northeast. Since this area is very vulnerable to incursions by people who might want to hunt, capture, or see timber rattlesnakes in the wild, I am compelled to refrain from mentioning the counties by name. Stechert has worked in one of the counties extensively. It's not surprising that three of his four longest timber rattlers have come from dens in that county. They ranged from fifty-four to sixty inches.[5] To illustrate how outstanding his five-foot-long snake was, Stechert tells people that he had actually given up hope of ever finding a sixty-inch-long timber rattler. After several decades in the field he had come to think of capturing a five-foot snake as an unattainable goal. He emphasizes that any timber rattler over forty-eight inches and sixteen hundred grams, approximately three and a half pounds, is a big snake. His five-foot-long snake weighed close to five and a half pounds.

In Stechert's opinion the Reeds' best snake might well have been close to five feet. Although snakes of this length are extremely rare, they are not out of the realm of possibility. He highly doubts that Art Moore's supposed seventy-inch-long snake was anywhere near that length, explaining that rattlesnakes can be stretched considerably after death, which would throw off the accuracy of measurements. W. S. Brown emphasizes this too. As far as the Galick snake being over six feet and possibly close to seven feet long, Stechert dismisses this as something akin to sheer fantasy.

In 1929, the widely read and respected herpetologist Raymond L. Ditmars reported a timber rattlesnake measuring six feet two inches from the town of Sheffield, in the Berkshire Mountains of western Massachusetts.[6] According to Stechert, W. H. Martin did some research to confirm this record and found not one but two possible problems with it. First, Mr. Ditmars assigned two different lengths to the snake at two different times in his career. In 1946, almost twenty years after his 1929 report on the snake's length, he referred to it as being only six feet long.[7]

In addition to this discrepancy, it is likely that Ditmars included the rattle when he measured the snake, because it was his habit to do so. Including rattle length when measuring snakes is not accepted by today's standards. It is comparable to including exceptionally long fingernails in the measurement of an arm. Without its rattle, which could very well have been three inches long, Ditmars's snake was probably somewhere between sixty-nine and seventy-one inches long, which still makes it the longest timber rattlesnake on record, according to Stechert. The longest modern-day timber rattler on record appears to be a snake captured in the last few years by Dr. Bruce Kingsbury on state forestland in Brown County, Indiana. This snake is reported to have been sixty-three inches long, excluding its rattle.

When one looks at the facts, it's clear that any timber rattlesnake between four and five feet in length is a large timber rattler and that any snake over five feet is an extreme rarity not only in the northeast but anywhere in their entire range. Did the well-known bounty hunters I interviewed slightly exaggerate the lengths of their longest-ever timber rattlers? It's possible that they did. Since almost everybody exaggerates length and size for effect, it would be foolish to think that they were incapable of doing this too. It's also possible that the bounty hunters told me what they thought to be the truth about the longest timber rattlers of their careers. In the case of Lester and Carol Reed there are two points to consider that could have thrown off the accuracy of their measurements significantly. The first is rattle length. The rattles on large timber rattlesnakes, even in the wild where unbroken rattles are somewhat uncommon, can be two inches long and longer. The second point has to do with the fact that dead rattlesnakes tend to stretch in length from dislocation of the vertebrae when they are lifted up off the ground. Their big snake may well have measured sixty-three inches, but subtracting two-plus inches for its rattle and an inch for the stretching that could have occurred in the first hour after its death would mean that the snake was actually closer to five feet, and probably a little less than that. If it was in fact that long, it was an incredibly fine timber rattler in the mid-1960s, and certainly one that Randy Stechert, or W. S. Brown, or W. H. Martin would love to have measured and weighed.

The same two factors that undoubtedly reduced the length of Lester and Carol Reed's big snake would also have reduced the length of Art Moore's supposed seventy-inch-long timber rattler: subtracting the length of the snake's rattle, and factoring in the dislocation of its vertebrae. A difference between Moore's big rattlesnake and the Reeds' may well have stemmed from the fact that Moore's snake hung for several days before it was measured. When rattlesnakes are hung up for several days, they can stretch up to three inches and more due to major dislocations of their vertebrae and the elasticity of their skins. In reality, Moore's longest timber rattlesnake may well have been an inch or two over five feet, making it an exceptional snake, but it was probably nowhere near seventy inches long.

I'll give the Galick brothers a little benefit of the doubt about their giant timber rattlesnake. It undoubtedly was a very special snake. Otherwise, they wouldn't have been so emphatic about its being over six feet long when they described it to me in their eighties. It's a human tendency to look at a relatively long snake, especially when it's in its habitat, and think it looks a lot longer than it actually is. As bounty hunters in Vermont, the brothers may not have brought the snake down from the den after shooting it, but only its head and rattle, as these were the necessary parts of a timber rattlesnake that had to be brought to town clerks in Vermont in order to collect the rattlesnake bounty.

The reason why long timber rattlers often look abnormally skinny and long in photographs is that they have been hung up for two or three days. The photograph of the two big rattlers being held up by Bill Galick is a prime example of this kind of stretching (see fig. 6).[8] Skins removed from dead timber rattlers can stretch over a foot in the skinning and drying process, accounting for the timber rattlesnake skins of five feet and more on so many walls throughout rural America.[9] It is pretty safe to assume that any such skins have come from snakes that were well under five feet in length when alive.

It's not hard to see why the Vertebrate Zoology Curator of the New York State Museum in Albany was willing to pay a reward of five thousand dollars in the 1960s for a timber rattlesnake over six feet long. Dr. Riley knew that the longest timber rattler on record at that point was the six-foot Berkshire County, Massachusetts snake discovered by Raymond Ditmars. If Art Moore had been able to capture such a snake alive, and if it held up in a comparison of measurements, the museum would have had the longest timber rattlesnake ever found in the wild and it would have attracted droves of herpetologists and curiosity seekers. Even behind glass, a timber rattlesnake this large would be frightening to encounter for most people.

Why is it that, as a species, we humans are so fascinated by long and giant snakes?[10] Why is it that a six-foot, or better still, a seven-foot timber rattler makes for a more interesting conversation or story than a mere five-footer? It's because we're not only fascinated by big snakes, we're frightened by them, and this fear probably goes back to the dawn of man. There's a strong argument that mankind has an innate fear of snakes. Other primates have exhibited a similar fear in laboratory tests. A seven-to-eight-foot timber rattler makes a better story than a five-footer because it scares us more, and oddly enough, man is attracted to things he fears. I recall, for example, conversations I overheard in the long ticket line outside one of the theaters in Rutland, Vermont, on the opening night of Steven Spielberg's famous movie Jaws. Many people were talking about being scared half to death, and that's what seemed to be attracting them more than anything else. I rest my case.

CHAPTER 3 : WHEN A TIMBER RATTLER BITES

When a timber rattlesnake envenomates a prey animal or human being, an extremely complex and fast-acting chain of biochemical events takes place. At least this is the norm, but timber rattlers, along with copperheads and cottonmouths, are capable of delivering so called dry bites in which they do not inject any venom at all, from 15% to 25% of the time.[1] In other words, these pitvipers are capable of metering the amount of venom they inject into their bites. I will attempt to explain in simple language from a general medical standpoint the various effects, ranging from tolerable (or no discomfort at all) to death, that people can experience after being bitten by one of these snakes. I'll also review what to do and what not to do after being envenomated.

Timber rattlesnake venom is composed of a variety of proteins, some enzymatic and others nonenzymatic, smaller in size, and referred to as polypeptides. Some of these proteins may have neurotoxic effects on the nervous system, while others have multiple hemotoxic effects on the blood and its components, as well as cytotoxic effects, or effects on cells. Simply put, a person gets a mixed bag of biochemical toxins which lead to a mixed bag of physiological effects after being envenomated by a timber rattlesnake. Although it's a formidable challenge to totally grasp these effects and what causes them, one doesn't need a Ph.D. in physiology or chemistry to gain a good basic understanding of them. There are a few terms and concepts with which one needs to become familiar, but before long, everything, even a worst-case post-bite scenario, falls into place and makes sense.

Anyone looking at a photograph of a timber rattlesnake bite victim who had been envenomated by a bite on one of his fingers and hadn't received medical help until four days later would immediately realize how serious a bite from one of these snakes can be. The victim's hand and possibly forearm and upper arm would most likely be incredibly swollen and also covered with unsightly blebs, or elevated fluid-blisters, of varying heights and in various shades of red and purple. He might also have patches of black skin, or dead tissue, intermingled with the blebs.[2] To understand how these conditions come about, we need to look at some of the components in the snakes' venom and how they work.

Swelling and blebs are caused by the shifting of blood fluid primarily from the vascular system into areas such as muscles, spaces between cells called intercellular spaces, cells themselves, and areas just below the skin. Collectively, these areas are referred to as extravascular compartments. The movement of blood fluid across vessel walls, or fluid shift, as it is known medically, is largely triggered by specific enzymatic proteins in the venom which attack cells and capillary walls and make them porous or leaky. Other enzymatic proteins in the venom target the blood directly and adversely affect its ability to clot. Platelets are small disk-like components of the blood that, in connection with platelet activating substances, promote coagulation and clotting. Medicine often makes references to platelet plugs and platelet plugging. One unique timber rattlesnake venom component, called crotalocytin, makes platelets aggregate and become nonfunctional, resulting in an actual lowering of the number of functional platelets necessary to form plugs in the vascular walls and reduce fluid shifts. Thus, it is the combination of capillary porosity and platelet malfunction that causes timber rattlesnake bite victims to experience fluid shift and swelling. Fortunately there are antivenoms, historically called antivenins, that can neutralize the venom components in vessels and blood responsible for fluid shift and swelling, and the sooner these antivenoms can be administered, the better. Every hour without antivenom is critical for a badly envenomated victim.

Black, necrotic areas of skin, which are often intermingled with a victim's blebs, are caused by certain enzymes in the venom that have the ability to destroy tissue. Tissue destruction gives timber rattlesnakes an advantage in the digestion of their prey. People who have been seriously envenomated by these snakes often experience some dead tissue, especially in the immediate area of their bites, which has to be cut away or debrided afterward. The key to avoiding tissue loss is to get to a hospital quickly. Findlay Russell, a physician who is arguably the best known venomous snake envenomation expert in the country, has found that tissue necrosis can be largely avoided in rattlesnake bite victims who receive the right amount of antivenom within two hours of being bitten.[3]

Many people seem to have the notion that timber rattlesnake envenomation is much more lethal to human beings than it actually is. W. S. Brown has frequently been asked, throughout his long career as a teacher and lecturer, how long a person can live after being bitten by a timber rattler. The first thing he points out is that virtually nobody dies nowadays from timber rattlesnake envenomations. As long as bite victims receive proper medical treatment soon enough, any complications that could, on a rare occasion, lead to death are pretty much arrested. He also points out that if a person does nothing at all after being bitten and receives no medical treatment whatsoever, the chances of that person dying would be somewhere in the neighborhood of 2% or less.

Brown goes on to explain that the rare individuals who die from timber rattler bites are often people who experience an adverse allergic response to the venom itself. Medical treatment doesn't always bring these individuals around. Their symptoms develop very quickly, and this anaphylaxis situation can be extremely life threatening. Under these circumstances the administration of adrenalin and antivenom can make the difference between life and death. The bottom line is to get to a hospital fast. If you're a person who is allergic to the venom, your life could depend on it. If you are not allergic, and therefore most likely a member of the general population at a low risk of dying from a timber rattlesnake bite, the advantage of receiving antivenom as quickly as possible is that a number of the symptoms commonly associated with timber rattlesnake envenomations such as tremendous swelling, pain, blood coagulation problems, tissue necrosis, nerve damage, nausea, and headache can be minimized and sometimes avoided completely.[4]

Bob Fritsch, a timber rattlesnake field worker in Connecticut during 1986, is a classic example of an individual who almost died after being envenomated by a timber rattlesnake, as a result of an allergic reaction to the snake's venom. I describe Fritsch's near-death experience later in this chapter. Fritsch was very thorough in giving me the details of what transpired after his bite, and the various aftereffects he endured.[5] He wasn't able, however, to explain the precise chain of medical events that put him so close to death, but Dr. Daniel E. Keyler, as a pharmacologist and toxicologist with the University of Minnesota at the Hennepin County Medical Center, was able to fill in the blanks in Fritsch's account.[6] The main culprit that proved nearly fatal for Bob was his loss of circulating blood volume because of fluid shift, resulting in a profound drop in blood pressure.[7]

I have discussed how fluid shift through perforated capillaries causes tremendous swelling in timber rattler bite victims, but this swelling is not in itself life-threatening. It's an indicator, rather, especially in extreme cases, of serious medical complications still to come. Adult human beings have approximately five liters of blood circulating continually throughout their bodies, pumped by the heart through a complex network of arteries, capillaries, and veins. The main function of this systemic circulation is to provide a constant fresh supply of oxygen, via the lungs, to the organs and muscles of the body. Fluid shift can lower arterial blood pressure to the point where the heart alone, no matter how strong it might be, has difficulty pumping a person's diminished circulating blood volume through his or her vascular system at a normal rate of speed. When this happens, the delivery of life-sustaining oxygen to the person's muscles and organs is reduced. Most adults understand the importance of maintaining a healthy level of blood pressure. High blood pressure can lead to an array of medical problems such as strokes, aneurisms, and heart failures. Whereas high blood pressure is a common topic of conversation, low blood pressure and its consequences are seldom thought about or discussed. Arterial blood pressure is created by the pressure that the walls of the arteries exert on the blood that flows within them. It generally stays in a normal range when there is enough blood in the arteries to create the right amount of pressure. It will begin to drop, however, if one's circulating blood volume decreases, a condition known as hypovolemia. In summary, if you are envenomated by a timber rattlesnake, and a subsequent fluid shift causes a precipitous drop in your blood pressure, you will have difficulty pumping the remaining blood in your vascular system, and the consequent oxygen deprivation would make your medical situation dire.

Through fluid shift, a timber rattlesnake bite victim could easily lose two or more liters of his or her circulating blood volume, which would lead to a dangerous drop in blood pressure. In Bob Fritsch's case, just two hours after his envenomation, his arterial blood pressure was an incredible forty-four over what is medically referred to as palpable, a condition in which the pulse can barely be felt. This was a far cry from his normal and healthy readings of one-hundred-and-twenty over eighty. His adverse symptoms were galloping because, in addition to the toxic effects of the venom, he was having an allergic response to the venom. Having experienced a significant amount of fluid shift into the compartments of his forearm, he was in what is medically known as hypovolemic (low blood volume) shock.

A question arises: How low can blood pressure go before it would cause a person's death? Dr. Keyler explained that any systolic number (the upper number in a blood pressure reading) in the forties would be considered a grave medical situation. He added that a systolic number lower than forty would almost certainly bring about cardiac arrest and death if untreated. Why? It's because the heart is a very oxygen demanding muscle. It only functions well when it continually receives the necessary volume of blood carrying fresh oxygen. In advanced hypovolemic shock, a person's heart becomes oxygen deprived and simply stops beating because the greatly reduced circulating blood volume is moving too slowly in the vascular system.

Hypovolemic shock is not the only way that a person could die from a timber rattlesnake envenomation. Dr. Keyler explained that some people might leak blood fluid so profusely into their lungs after being bitten by one of these snakes that it could cause them to drown from their own body fluids. He pointed out that this would be highly unlikely but definitely possible. Generally speaking, when people are fluid-shifting that badly, they're more likely to die from the classic symptoms of hypovolemic shock—decreased circulating blood volume, low blood pressure, and lack of oxygen to the heart—before drowning. Keyler added that some pooling of blood fluid in the lung tissues is a common symptom associated with hypovolemic shock, and in a badly envenomated timber rattlesnake bite victim the lungs can become the target organ for the pooling of fluid.

Full-blown anaphylaxis (often referred to as anaphylactic shock), which results from an intense allergic reaction to the venom, has some distinct parallels with hypovolemic shock. In both conditions, there's some degree of fluid shift or loss of blood fluid through perforated blood vessels, along with a marked drop in blood pressure, a reduced rate of flow of one's circulating blood volume, and some pooling of blood fluid in the lungs and other organs. Where anaphylactic shock differs from hypovolemic shock is the manner and speed in which it can prove fatal. In acute anaphylaxis, a person goes into cardiac arrest from swelling in the upper trachea, an inability to breathe, and oxygen deprivation. This can kill a person in a matter of minutes, whereas hypovolemic shock from a rattlesnake envenomation typically takes the better part of a day or longer to cause a person's demise.[8]

It's easy to wonder why more people don't die after being bitten by timber rattlesnakes since envenomations by these snakes can create so many serious internal problems. The answer lies not only in the fact that the general population tends to experience less severe medical complications than those experienced by the rare individuals like Bob Fritsch who are allergic to the snake's venom, but in two other facts as well. First, it so happens that the more dangerous components in timber rattlesnake venom tend to disperse gradually throughout one's body via the lymphatic system, and most people are able to receive antivenom relatively soon after being bitten. Second, it's important to bear in mind that timber rattlesnake venom (as is the case with all rattlesnake venoms) evolved to quickly immobilize and kill small prey animals. It did not evolve to deal with large mammals, including human beings.

When animals such as white-footed mice or chipmunks are envenomated by a timber rattler, they begin to bleed profusely throughout their bodies. As previously described, they attempt to run away but aren't able to go very far before becoming immobile, kicking spasmodically, and dying. It isn't fluid shift and hypovolemic shock that so rapidly kill them but rather certain components in the rattlesnake's venom that cause severe, rapid coagulation complications, leading to heart failure and death within seconds. Dr. Keyler once witnessed a mouse, which he fed to a timber rattlesnake in a laboratory, die ten seconds after being bitten. That was an exceptionally fast death. The only way a small mammal could die any faster from a rattlesnake bite would be if the snake's fangs penetrated either its heart or lungs directly, causing near instant death.

As discussed, human beings almost never succumb to timber rattlesnake envenomations and are able to minimize the unpleasant effects from these poisonings when they are able to receive medical treatment and antivenom quickly enough. Until recently, the most commonly used antivenom available to rattlesnake bite victims was produced by Wyeth Laboratories. This traditional antivenom was the serum (clear liquid blood component) obtained from horses that were repeatedly injected with small doses of pitviper venoms, mostly obtained from various species of North American rattlesnakes, and also from the South American fer-de-lance. To counteract the effects of their envenomations, the injected horses' immune systems produced antibodies in their blood that gave them immunity to the venom components. When Wyeth antivenom, or antiserum, is introduced intravenously to a rattlesnake bite victim, its equine (horse) antibodies, which are proteins, bind to the venom proteins in the blood of the victim and neutralize them.

The downside of Wyeth antivenom is that treated patients may develop an allergic reaction to the horse-based antibodies because they are recognized as foreign to the person's immune system. The norm for people who have developed an allergy to Wyeth antivenom is to experience what is known as serum sickness up to fourteen to twenty days after being treated. This is called a delayed allergic response. The symptoms of serum sickness can be almost negligible in some cases. I am aware of an individual, named Milan Fiske, who received seventeen vials of Wyeth antivenom after he was bitten by a timber rattlesnake in the eastern foothills of the Adirondacks, and developed a very mild case of serum sickness afterward. Mr. Fiske was fortunate. All he experienced were bouts of lassitude and weakness that lasted for several weeks.[9] In addition to feeling weak, people with serum sickness often experience feelings of nausea along with aches in their joints. In one of many conversations with Dr. Keyler, I learned that flu-like symptoms are commonly associated with serum sickness.

Many people experience much worse allergic reactions than Mr. Fiske's to the antivenom that is commonly referred to as "Wyeth." In some cases, people break into itching and hives as soon as they begin to receive this antivenom. This happened to W. S. Brown while being treated with Wyeth in September of 2003. In more serious cases, people may develop rapid heartbeat, sweating, excessively rapid respiration, or difficulty in breathing. In an extreme case, a person could develop acute anaphylaxis, leading to a swelling and blocking of the upper airways. When any of these symptoms occur, the physician treating the bite victim will usually back off the speed, or rate, with which the Wyeth is being administered, or will stop the infusion for five or ten minutes, then start it again very slowly. It is also common to give the patient an intravenous antihistamine such as Benadryl™ in these situations. Generally speaking, any serious allergic reactions can potentially be counteracted. What's critical is that a patient is monitored closely during an infusion of Wyeth, since people have been known to develop serious allergic responses to the antivenom in the first fifteen or so minutes.

One of the first things that is done medically when a timber rattlesnake or other pitviper bite victim comes into an emergency room is a skin test to see if the person is allergic to Wyeth antivenom. Many hospitals are still using Wyeth even though CroFab™, a newer antivenom with far fewer allergic complications and not needing skin testing, arrived on the market in 2001. In a Wyeth antivenom allergy skin test, a tiny amount of the antivenom or a dilution of normal horse serum is injected under the skin. If a wheal and flair, or swelling and red ring, appears on the skin, the medical personnel immediately know that the individual is potentially allergic to the substance. In these cases, if the medical personnel feel that the patient's bite is serious enough to necessitate the use of Wyeth, they begin to infuse it very slowly and stand by ready to use an antihistamine and adrenalin (epinephrine) if and when any allergic symptoms manifest themselves. It's a common practice in cases such as these to have epinephrine and Benadryl™ "piggy-backed" onto the main i.v. line and thus ready for immediate use. Patients who are known to be allergic to horse serum or have tested positive to the Wyeth skin test have to be monitored extremely carefully and are typically placed in an ICU (intensive care unit).[10]

Fortunately, the days of developing severe allergic reactions to rattlesnake antivenom are dwindling and will soon become a phenomenon of the past. This in large part is due to the gradual disappearance of Wyeth antivenom in hospitals throughout the country, and the increasingly widespread use of CroFab™. In the near future, Wyeth antivenom will probably become totally obsolete, since it is not currently being manufactured and has a shelf life of only about five years. The only hospitals still using this antivenom are ones that have it in stock. Dr. Keyler told me that people are experiencing far fewer and less severe symptoms with CroFab™, and that he personally knows of no cases in which this new antivenom has triggered anaphylactic shock. Although patients treated with CroFab™ sometimes have a recurrence of local effects and blood platelet abnormalities requiring repeated follow-up dosing of CroFab™, the important secret to its advantages for patients lies in the way in which it is manufactured.

Where Wyeth antivenom consists of antibodies derived from horses, CroFab™ consists of antibodies derived from sheep. CroFab™ is made by injecting sheep with small doses of pitviper venoms from various species of North American rattlesnakes, and also the cottonmouth. Subsequently, the sheep's immune systems produce antibodies which are harvested from the sheep's blood and chemically broken down into antibody components that are one-third the size of normal mammalian antibodies. Wyeth and CroFab™ antibodies are both viewed as foreign proteins when they enter a person's body. The difference between the two antivenoms is that a person's immune system and body can deal with CroFab™ much more easily because of its smaller antibody size.

Although the vast majority of people survive after being envenomated by a timber rattlesnake, most people can expect to experience at least some unpleasant symptoms. Virtually nobody gets off scot-free. If nothing else, the majority of envenomations cause tremendous, immediate pain and swelling around their bite sites.[11] In addition to this, most victims go on to experience some additional unpleasant symptoms, such as those already discussed, while being treated for their bites, caused by components in the venom or by an allergic reaction to the antivenom. While a full recovery from these symptoms is the norm, body reactions often leave those who have received serious envenomations with some permanent impairments. They serve as grim reminders to these individuals for the rest of their lives to beware of timber rattlesnakes and their potential danger. This has certainly been the case with W. S. Brown and Bob Fritsch.

It now appears as if Brown may not regain the normal sensation of touch or the strength in his left thumb, after receiving a serious bite on the digit in the summer of 2003. He has trouble performing many minor tasks that he used to take for granted, such as buttoning his shirts, tying his shoes, using dental floss, or opening packaging such as bags of potato chips. When he touches anything with his left thumb, the thumb doesn't feel right. If he touches hard surfaces even lightly, he can get a painful sensation in the thumb. He kiddingly makes references to his "experimental thumb" and contrasts it to his normal "control" thumb. A few months after receiving this bite he said, "It's a lucky thing that snake didn't bite me on my right thumb." Since Brown is right-handed, it was easy to catch his meaning.[12]

Bob Fritsch does a lot of carpentry work and is physically active in a number of ways. As part of his part-time job at a local hardware store, he unloads the supply trucks that come to the store. While unloading, he is always aware of the decreased strength in his left forearm due to an invasive surgical procedure called a fasciotomy that was performed on his forearm several hours after his bite. In his daily life, he is also frequently aware of a lack of dexterity in his left hand. Not only is his middle finger, the digit that was bitten in 1986, less dexterous, but so are the two fingers on either side of it. According to him, he has lost the coordination between those three fingers and is slightly disabled because of it. In addition to this, his bitten finger is slightly numb all the time and extremely sensitive in cold weather. The reason he is experiencing problems with his fingers probably stems from the fact that he sustained some permanent damage to his nerves because of the tremendous compartment pressure that developed in his hand and arm after his bite. Additional damage to his nerves may be traceable to the fasciotomy he underwent. All in all, though, he seems to be faring better than Brown in the area of long-lasting bite reminders. He's been lucky in this regard, considering the seriousness of his case.[13]

Even though it's probably safe to say that no two timber rattlesnake bite cases are identical, everyone who is bitten by one of these snakes is likely to have one of five different post-bite experiences. The first experience is receiving a dry bite and not being envenomated at all. The second is being bitten and envenomated (in some cases seriously) but not seeking out medical help for any one of a number of reasons including embarrassment, professional pride, religious faith, and the exorbitant cost of treating snakebite cases in hospitals.[14] I refer to those who handle a bite this way as "tough-it-outers."[15] The third experience is being bitten and envenomated, then going to a hospital, receiving antivenom, and being released when the symptoms are brought under control. The fourth experience is being bitten and envenomated, and nearly dying afterward. The fifth experience is dying after being bitten and envenomated. Death pursuant to timber rattlesnake envenomation is so uncommon that I rarely discuss it. I am not aware of any reliable, detailed accounts of individuals who have met this fate.[16]

Dr. Keyler is an example of an individual who experienced a dry bite from a timber rattlesnake. When Keyler has time away from his busy career as a researcher and full professor at the University of Minnesota Medical School, he is likely to be found in the field doing timber rattlesnake research under grants from the states of Minnesota and Wisconsin. The underlying goal of his research, whether it focuses on reproduction, distribution, or some other area, is the preservation of the species in the upper midwest. On the sweltering afternoon of August 13, 1995, Keyler and his fellow field worker, Barney Oldfield, were on a field trip in Winona County, Minnesota, when something completely unexpected took place.[17]

At about two o'clock Keyler discovered a small adult timber rattler subtly blended in under some prairie grass on the slope of a bluff overlooking the Mississippi River. He carefully drew the snake out of its concealment with his hook, placed it on top of the grass, and hollered to Oldfield to come over and see it. With his hook in his right hand, he gently pinned the snake down by its neck on the long dry grass. Everything was fine up to this point. Then, with the snake well secured, he reached forward to pick it up behind its head with the thumb, index, and middle finger of his left hand. It's a maneuver he had executed countless times before with one hand or the other, only this time his approach was slightly different. Normally he stands behind snakes before picking them up in this manner. On this particular afternoon he was standing in front, probably because of the steepness of the slope and the fact that he was downhill from the snake when he discovered it.

Keyler was just making contact with the snake after reaching forward from in front of and to the right side of its head when the snake suddenly opened its mouth, swung its fangs to the right, and nailed him on his thumb. He had the presence of mind to rapidly bag the snake before sitting down to examine his wound. Two distinct fang punctures were both bleeding. Things didn't look good. He grabbed his Sawyer Extractor™ as quickly as he could and placed the largest cup over both punctures. It readily filled up with fluid, so to be on the safe side he applied the cup two more times. During this process Keyler realized that he was experiencing a very slight burn in the area of his punctures but no pain per se. Oldfield wanted to leave immediately and get to medical assistance, but Keyler urged him to sit down and wait to see what happened. He admitted to me that this may not have been the smartest action to take, but he was pretty confident that he hadn't been envenomated. After an hour of sitting, he got up and suggested that they go on to another site. Upon returning home later in the day, he thoroughly washed his wound with soap and water, and his life quickly returned to normal. He'd been in the lucky 15% to 25% of timber rattlesnake bite victims who receive a dry bite.

When it comes to toughing-out a timber rattlesnake bite and not bothering to seek out any medical assistance afterward, we have an excellent first-hand account by a man whose name was Fred Stiles. Stiles was a dairy farmer in northeastern New York in the 1930s when the event took place. He detailed the experience in a book he wrote fifty years later.[18] At the time of his bite, he was keeping a small herd of cows and heifers on his farm in Washington County. One morning, a Holstein bull that normally ran with his cows came into the barn acting very sick. It was early and not yet full daylight. Stiles left the barn and went out into his pasture to round up some straggling cows. Following a narrow pathway through some overhanging bushes, he saw what he thought to be a large pile of cow manure on the path. He tried to step over it but didn't quite make it. The manure he thought he was landing on turned out to be a very large timber rattlesnake. Instantly the snake sank its fangs into Stiles's leg right below his knee. He was in pain immediately and had to strike the snake's head to make it let go. He then went to a little nearby stream, washed the bite area as well as he could, and went back to his chores. Later that morning the bull died.

What's remarkable about Stiles's experience is that he apparently didn't even consider going to a doctor. He explained that he had too much work to do and nobody else to do it. His leg was swollen and blue in color for two weeks after his bite and, not surprisingly, he felt terrible. Another interesting aspect of this account is that when Stiles studied the snake he found the hoof print of the bull on its hide. It was then he realized that the big rattlesnake had bitten the bull before it had bitten him. Had the big snake injected as much venom into Stiles as it had injected into the bull, would he have died? He probably would have been completely incapacitated for a number of days, and ultimately lost some flesh due to necrosis, but the odds are that he would have survived with some degree of permanent impairment.

A few days after this incident, an old rattlesnake hunter, whose name was Bill Smith, came by to see the skin, which Stiles had removed from the snake, and offered to buy the rattle from it. He said that this rattlesnake was larger than any he had ever killed and that he had killed many of them in his time. The skin supposedly measured six feet, two inches including its rattle, which had "sixteen segments and a button." There's no question that this snake was a formidable specimen, but it's unlikely that its live length was anywhere near six feet. As pointed out in the previous chapter, fresh skins stretch very easily. All in all, it seems pretty clear that Fred Stiles was bitten by a much bigger timber rattler than would commonly be seen or captured in Washington County today. It may well have been a genuine five-footer.

In the twenty-five years that W. S. Brown has been involved with his timber rattlesnake field study, he has had two serious bites that required hospitalization. Considering the fact that Brown has captured and released over four thousand snakes, it could be argued that it's remarkable he hasn't been bitten and hospitalized more often. Nobody was amazed when Evel Knievel had accidents with his motorcycle. Knievel's accidents were an inevitable part of his career as a daredevil. Similarly, timber rattlesnake bites are bound to occur in Brown's occupation, although he can hardly be described as a daredevil. He is as careful as possible, but eventually, given the sheer volume of the snakes he restrains, handles, marks, and examines, he's likely to be bitten. Of the two bites that required hospitalization, one in particular strikes me as being a typical timber rattlesnake bite experience. It occurred one afternoon in June 1985, about seven years into his field study.[19]

Brown had just parked near his lab facility in the Adirondack foothills. He had a bag containing four big rattlers that he had captured three days earlier, brought to his lab to mark and measure, and was about to take back into the woods to release. He was lowering the bag into his backpack, a normal procedure he had performed many times previously, when his left hand accidentally brushed against the top portion of the bag just below the knot he had tied in it.

He felt what he described as a pricking sensation and looked down to see a slashing wound in the fleshy area of his left hand at the base of his little finger. For several seconds he looked at the wound, not wanting to believe he'd been bitten.

Then, realizing what had happened, he frantically tore open his pack for his snakebite kit. In his near panic (he had never been bitten before), he fumbled with the kit's constriction band. Unable to secure it tightly around his lower arm, he gave up with it and, using the kit's suction pump, tried to suck the venom out of his wound instead. By now his wound was bleeding profusely, and his hand was throbbing in the area of his bite. He tried to locate a friend who had an office nearby, but he couldn't find him initially. When he did locate him, he said he'd been bitten by a rattlesnake. His friend immediately called the local rescue squad. While the ambulance was on its way, Brown had the presence of mind to call his home and keep sucking blood out of his wound. He also asked his friend's wife to help him remove his wedding band, which was feeling very tight on his finger. Anticipating that major swelling in his left hand was imminent, this was a wise decision.

Before long, approximately fifteen minutes post-bite, Brown was on his way to the Glens Falls Hospital. In the forty or so minutes it took to get there, he kept applying suction to his wound and emptying the suction pump onto bloody towels. (The ambulance crew members deferred to Brown's knowledge of the subject.) Ten minutes into the ride he began to feel some signs of paresthesia, or a burning and tingling sensation of the skin, in his face and tongue. Paresthesia is common after timber rattlesnake envenomations. Swelling was also beginning to develop on the top of his hand. By the time he arrived at the hospital, the constriction band, which his friend had been able to apply, was moved halfway up his arm to his elbow to keep ahead of the swelling.

When Brown arrived at the emergency room, he was relieved to find that the hospital had a copy of Findlay Russell's *Snake Venom Poisoning*. On the other hand, the hospital had only three vials of Wyeth antivenom on hand. Blood was drawn and a skin test administered to see if Brown was allergic to antivenom. It turned out he wasn't, which was a plus. While in the emergency room, his hand became more painful. It was now swollen to the point where he couldn't move his fingers. In addition to this, swelling was moving up his forearm and approaching his elbow. He was transferred to a small room across from the emergency room, and his first drip of antivenom was started at about five-thirty in the afternoon, roughly two hours after his bite. During this time, he called and reached his wife, who had just arrived home. He explained his situation and asked her to bring up at least six units of antivenom from his lab at Skidmore College.

It was about seven–thirty in the evening when she arrived at the hospital with the additional antivenom he had requested. Brown had already been given two vials of Wyeth antivenom and they were starting on the third. By now, approximately four hours post-bite, his arm was swollen up to his shoulder. One young doctor attending Brown suggested doing a fasciotomy to relieve the swelling. Having read Dr. Russell's warnings about this procedure, Brown turned him down flatly, even though his arm was severely swollen. He said that the operation would be damaging and unnecessary, and recommended that the doctor should read Russell's opinion of the procedure in *Snake Venom Poisoning*.[20] The doctor left, read Russell's section on fasciotomy, and returned with a different decision; instead of performing a fasciotomy, he would continue the antivenom while monitoring Brown's symptoms closely throughout the evening. In a fasciotomy, long cuts parallel to the axis of the limb are made through the fascia (the sheaths surrounding muscles) in order to release excessive fluid pressure on the muscles. Most doctors and surgeons who treat and are familiar with rattlesnake envenomations try to avoid fasciotomies because of the damage they can inflict. For the most part they only perform the procedure when a patient's compartment pressure is high enough to start causing major muscle damage, and when they feel they have no alternative.

After receiving the third vial of Wyeth, Brown was moved to the ICU where he received four more vials, totaling seven in all. Five to ten vials of antivenom fall within the normal guidelines for what are considered to be moderate cases of rattlesnake envenomation. Despite having swelling all the way up to his shoulder, Brown's was a moderate envenomation.[21] Although some degree of fluid shift had taken place, causing the swelling and elevated muscle compartment pressure in his forearm, his blood pressure was not plummeting. Therefore, hypovolemic shock, as previously discussed, had not developed, but his platelet count, the level of platelets in his blood, was of some concern to the team that was treating him.

Platelet counts normally run somewhere from 170,000 and 375,000/mm³. Brown's count was 89,000/mm³ in the emergency room and bottomed out at 27,000/mm³ in the ICU the day after his bite. In short, he was at risk of excessive internal bleeding. In his second full day in the ICU, his count increased slightly to 29,000/mm³. During his third day, it increased further to 52,000/mm³.

He was well on his way to recovery. For the first time since his bite, he felt much improved. Later that day he was moved out of the ICU into a regular room. The following day, his platelet count was up to 82,000/mm³. All other signs looked favorable, including his blood pressure and swelling, which by now was almost down to normal. Late in the morning of the fourth day his doctor came in and discharged him.

As mentioned earlier, Bob Fritsch is an example of my fourth category of timber rattlesnake bite victims, those who nearly die after their envenomations.[22] In the fall of 1986, he and a fellow herpetologist, Dr. Douglas Fraser, were on a timber rattlesnake field trip not far from Hartford, Connecticut, when Fritsch made a serious mistake that nearly cost him his life. He was attempting to lower a timber rattlesnake into a long-handled collecting bag, which he was holding in his left hand. The rattler was draped on the end of his long-handled snake hook, which he was holding in his right hand. This was a routine procedure he had performed on many previous occasions. Suddenly the unexpected happened. Fritsch lost his footing and, in an instinctive response to regain his balance, his left arm shot forward. This placed his hand perilously close to the snake, which moved up the handle of the snake hook and bit him on his left middle finger.

Perhaps it's a little unfair to say that he made a mistake. It's very easy to lose one's footing in timber rattlesnake habitat. It could be argued that his mistake was in using a snake hook instead of a pair of snake tongs to handle the snake in the first place. Why? While draped on a snake hook, a rattlesnake has freedom of movement. A safer way to lower a timber rattler into a collecting bag is to secure it firmly in its mid-section with a pair of long-handled tongs, thus greatly restricting its movement. Herpetologists who don't realize or remember this fact set themselves up for snakebites. Had Fritsch been using snake tongs rather than a snake hook that afternoon in 1986, the snake that bit him would never have gotten close enough to do so.

After Fritsch was bitten, he had a typical initial reaction. He couldn't believe what had happened. Then the realization set in. Within two minutes after his bite, his opposite hand, face, and feet were tingling, and adrenalin was surging throughout his body. Although Fritsch was a policeman at the time and not a wimp by any means, he freely admitted to me that panic began to overtake him. Instinctively he knew that he was in serious trouble. He was experiencing the first symptoms of a serious allergic reaction to the venom and the worst was yet to come.

If Fritsch had been alone, he would never have made it out of the woods. He would have died. As it was, Doug Fraser's presence and level headedness in the face of crisis saved his life. It also gave the medical community at large a classic example of a nearly fatal timber rattlesnake envenomation. Fraser could readily see that he had to get Fritsch out of the woods and to the hospital as soon as possible. The problem was that Fritsch was too weak and dizzy to climb up the steep hillside between their position and where Fraser had parked his truck. Little by little they advanced upwards, but Fritsch kept falling down and gasping for air. When it became apparent to Fraser that Fritsch was hindering him, he decided to leave him behind and run for help. Luckily for Fritsch, Fraser was as an avid runner and was able to cover ground quickly through the woods back to his pickup truck, which was parked about a half-hour walk away. Also lucky for Fritsch was the fact that Fraser was a knowledgeable herpetologist who could make good decisions under pressure.

Fraser knew full well that time was of the essence as he ran through the woods toward his pickup truck. He realized that he'd have to get the local rescue squad to Fritsch as quickly as possible, and that the route he and Fritsch had taken into the woods that day would not be fast enough. There had to be a better way. It occurred to him that the rescue personnel could get close to the location with a four-wheel-drive vehicle if they drove up through a clearing in the woods from the main road below. And this is exactly the way things played out. When Fraser arrived at his truck, he drove down to the main road to a house he and Fritsch knew near the clearing. Someone was at home and Fraser put in an immediate call to the local rescue squad. He explained what had happened, how serious the situation looked, and how urgently help was needed at the intersection of the clearing and the road he was calling from. He then had the presence of mind to put in calls to the Arizona Poison and Drug Information Center in Tucson, the Hartford Hospital, and Fritsch's wife, Janie. After his calls, he quickly rushed to the edge of the clearing and waited. Soon people from the local fire department, rescue squad, and ambulance began to arrive. News of Fraser's urgent call had traveled quickly. Within minutes he was on his way uphill with a small rescue team in four-wheel-drive pickup trucks.

When the team of rescuers reached Fritsch, who was well into the woods off the clearing, it was five minutes to four, an hour and ten minutes after his envenomation. His blood pressure was fine at 120 over 80, but his pulse was 85 beats per minute and thready, or weak.[23] It was clear from the way Fritsch looked, his labored breathing, his thready pulse, and Fraser's input on the severity of the situation that Hartford Hospital's emergency helicopter, better known as "Life Star," had to be summoned. They couldn't initially reach the helicopter because it was on another call, so they used the time to get Fritsch out to the clearing on a backboard with straps. More calls were made to "Life Star" before they were able to make contact. After that everything happened quickly. The helicopter came to the clearing, Fritsch was loaded on board, and he was in the emergency room at Hartford Hospital at four-thirty p.m.

A doctor named Jack Lynch and his team sprang into action. Fritsch remembers having a monstrous thirst when he got to the hospital, and his arm feeling as cold as ice even though it was warm to the touch, but these were the least of his problems.[24] He was in serious hypovolemic shock with his blood pressure at an incredible forty-four over palpable and his platelet count at zero. He'd been leaking blood fluid profusely from the vessels of his left forearm for some time, causing it to swell enormously. It was pretty apparent that he was experiencing an acute allergic response to the venom, because his symptoms were way beyond and far more serious than what would normally be expected in a victim a little under two hours post-bite. In addition to this, he had struggled to get enough air to breathe almost immediately after being bitten, which is a classic sign of anaphylaxis.

Fritsch tested negative to the antivenom skin test and was started on Wyeth almost immediately. All in all he would receive twenty vials of Wyeth in the next twelve hours.[25] He was also given a blood transfusion almost immediately in an attempt to raise his platelet count. Amazingly, he began to stabilize rapidly during his first few hours in the hospital. His blood pressure rose to nearly normal and so did his platelet level. He was no longer in danger of losing his life, but something else was seriously wrong. All during the evening Dr. Lynch and his team were monitoring the compartment pressure in Bob's left forearm, and it was getting higher and higher. Although Bob was going in and out of consciousness, he remembers how concerned Dr. Lynch and Dr. Don Kelly, an orthopedic surgeon, were about his forearm, and how conscientious Dr. Kelly was about trying to explain a possible fasciotomy to him.

Essentially, the two doctors were facing a dilemma. Should they wait to see if the antivenom would begin to lower the compartment pressure in Bob's forearm? Or should they perform a fasciotomy to relieve the pressure? The risk of waiting and not performing a fasciotomy was the possibility that the muscles in Fritsch's forearm could begin to die. When muscle compartment pressure is extremely high because of fluid shift, it can greatly reduce the flow of oxygenated blood to a muscle. Without sufficient oxygen, muscles die rather quickly, and once they're dead, there's no medical procedure to restore them. Dr. Lynch and Dr. Kelly feared that any significant loss of muscle in Bob's forearm might cost him his job in the local police department.

The risk of not waiting long enough to see if the antivenom would lower Bob's compartment pressure, and going ahead and performing a fasciotomy on his arm, was that the procedure itself could permanently weaken or damage his muscles, as well as damage his nerves. At about ten-thirty in the evening, as Dr. Lynch and his team were waiting to see if Fritsch's condition would improve or worsen, the compartment pressure in his forearm reached a point where it was seven to eight times higher than it would be under normal circumstances. The team's dilemma of indecision was over. It was now apparent to Dr. Lynch and Dr. Kelly that surgery was unavoidable. Fifteen minutes later Fritsch was in the operating room undergoing an extensive fasciotomy during which three deep, eleven-inch-long cuts were made in his forearm. Fritsch came through the operation well. His compartment pressure returned to normal, but his immune system wasn't quite finished with him yet. Although he had tested negative to the antivenom skin test when he arrived at the hospital, at some point while receiving twenty vials of Wyeth, he had apparently become allergic to it and begun to react to it. I asked him if he felt any of the symptoms of delayed serum sickness like those Milan Fiske had experienced. If he did to some extent, he doesn't remember. What he recalls is a nasty case of hives, which itched so badly he had trouble sleeping at night. Considering everything that Fritsch endured, I imagine he'll think twice before trying to bag a timber rattlesnake with a snake hook rather than tongs in the future.

HOW TO REACT; WHAT TO DO

Although most readers will never be bitten by a timber rattlesnake or come upon one in the wild, for those who might possibly be bitten by one of these snakes, and for those who are curious about the best way to respond and what first aid measures to take after being bitten, the following information may be useful. First and foremost, avoid panic. This is easier said than done, but it's important to bear in mind that the large majority of victims live after being bitten by a timber rattlesnake. Whether alone in the woods or with another person who has been bitten, panicking is counterproductive. A bite victim needs to take a few basic measures and attempt to get to the nearest hospital as quickly as possible. In order to meet these needs, rational thinking is essential.

After being bitten by a timber rattlesnake, you need to determine whether you have been envenomated or not. Bear in mind that timber rattlers are capable of delivering dry bites. Seeing obvious fang punctures doesn't necessarily mean that you have been envenomated. If you experience pain, swelling, and bleeding from the fang punctures almost immediately, you have been envenomated and need to take action. If you have a cell phone, call 911 before attempting any first aid. Explain what has happened, how far in the woods you are, and the location of your vehicle. You need to have an ambulance sent to you as soon as possible, and the hospital to which you will be going needs to be anticipating and preparing for a rattlesnake bite emergency. Calling 911 will accomplish both of these needs.

Before starting out of the woods it's very important to remove any rings or bracelets, if the bite was on your hand or arm.[26] This is to avoid any complications arising from the extensive swelling that is likely to follow, and is a procedure most experts in the field of venomous snakebite currently advocate. After this, your number-one objective is to get back to your vehicle as quickly and safely as possible. If you are panicking and unable to think rationally, you could easily fall and get hurt. You could also lose your mental compass and begin to walk in circles or in the wrong direction. Upon arriving back at your vehicle, go to the nearest phone and call 911 if you weren't able to make the call in the woods. You have done your part; now it's time to trust yourself to others.

Many people may wonder if there isn't more that can be done after being bitten and envenomated by a timber rattlesnake in the field. There are three different first aid measures that were recommended until quite recently for the treatment of pitviper envenomations: limb wrapping with tourniquets or constriction bands to slow down the spread of the venom through the lymph system, making cuts through or near fang punctures, and applying suction afterward. Limb wrapping is currently considered a highly questionable practice, and it is not routinely prescribed by most experts in the field of venomous snakebite poisoning from rattlesnake envenomations.[27] The reason that wrapping is not recommended in rattlesnake envenomations is that it can hold in place the components in the venom that are so damaging to tissue, and thus create a lot of tissue destruction. As for making cuts through or near fang punctures and applying suction afterward, there has been a total reversal in thinking over the last forty or so years, and these procedures are no longer recommended either. If anything, most experts in this field strongly recommend that these procedures should be avoided, since cutting can be very dangerous and suction is basically useless.[28]

There are several reasons why cutting can be dangerous. As previously discussed, various components in timber rattlesnake venom can cause significant bleeding. It is now known that most deaths from rattlesnake envenomations stem from a loss of circulating blood volume. Therefore, doing anything that might significantly increase your loss of blood is highly unadvisable. Another reason not to cut is to avoid serious infection. You already have venom in your system. Why take the risk of introducing any germs into your system as well? Germs introduced into the body by cutting deeply into fang punctures could spread rapidly throughout the body via perforated vascular walls. The primary reason to avoid cutting, however, is that nerves or major blood vessels could be severed and lead to either increased bleeding or complications that could lead to a loss of function, or even amputation, of toes or fingers in severe cases.[29] The reason why any attempt to suck timber rattlesnake venom out of fang punctures is largely futile has to do with the speed with which pitviper venom absorbs into the tissue surrounding bites. A bite victim has a very limited window of opportunity to respond before the venom is substantially absorbed. Once absorbed, it's as hard to remove as ink dripped on a wet sponge, according to one expert. Therefore, even though suction cups have been included in snake bite kits for many years, they are of little value to an individual who has been bitten by a timber rattlesnake, or any other pitviper.[30]

I have discussed with W. S. Brown the subject of first aid measures and his own response after being bitten on the thumb and middle finger of his left hand while attempting to bag a pregnant female with a snake hook in September 2003. Apparently, his hand started to swell so quickly that he wasn't able to remove his wedding ring. The only thing he did in terms of first aid was to fashion two constriction bands out of a section of bootlace and a section of lanyard material, and tie them around his left forearm in two separate places above his wrist. He did this even though he was fully aware of the current consensus among experts on the futility of limb wrapping after being envenomated by a pitviper. He was hoping to slow the spread of venom throughout his lymph system. After securing the bands snuggly around his forearm, he put in an emergency call with his two-way radio and began walking out of the woods. In about thirty minutes he arrived at his car, and soon afterward was on his way to the hospital in the local ambulance.

What I find interesting in Brown's description of his experience is the fact that he flirted with panic to some extent, even though he'd been bitten by timber rattlesnakes on previous occasions and knew how dangerous panicking could be. When he looked down at his left hand within a second after being bitten, it was already bleeding and burning with pain. Fifteen seconds later he began to feel stinging, tingling, and numbing sensations all over his face (the telltale signs of paresthesia). No wonder he started to panic. He feared that his bite was serious. It's probably safe to say that panic is a normal reaction after being bitten by a timber rattlesnake, especially when the bite victim is alone and deep within the woods. If these victims knew how best to react, that there are virtually no effective first aid measures to take other than removing rings and bracelets, and the all-important fact that almost nobody dies after being bitten by one of these snakes, it could help them immensely. It's this very knowledge, essentially, coupled with his experience, that gave Brown the ability to rise above his fears, act rationally, and get to the help he needed as quickly as possible in the summer of 2003. If you happen to be bitten and seriously envenomated by a timber rattlesnake while deep within the woods, a little key knowledge regarding these envenomations could prove to be your greatest asset.

PLATE 1. A 39″ female black morph timber rattler in defensive posture. Photo taken in Warren County, New York, on 6/7/1984 by Randy Stechert.

PLATE 2. A 48″ male yellow morph timber rattler in ambush posture. Photo taken in southern New York State 6/4/2004 by Randy Stechert.

PLATE 3. A 46″ male black morph timber rattler. Photo taken in southern New York State on 6/14/2003 by Jon Furman.

PLATE 4. A yellow morph timber rattler basking in front of its shelter rock. Photo taken in southern New York State on 5/1/2003 by Randy Stechert.

PLATE 5. A defensive yellow morph timber rattler. Photo taken in Washington County, New York, on 10/10/2002 by Jon Furman.

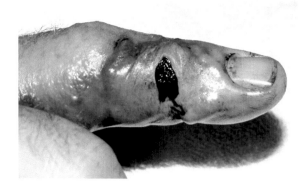

PLATE 6. An untreated timber rattlesnake bite victim four hours post-bite. Photo taken in an emergency department in June 2005 by Dr. Daniel E. Keyler.

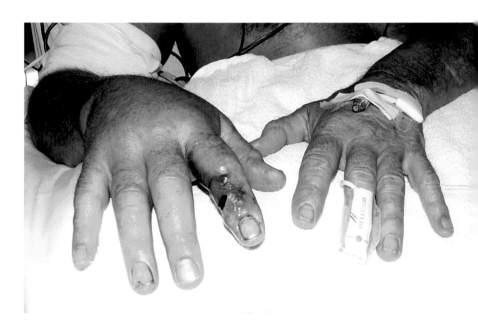

PLATE 7. The same victim twenty-two hours later in the medical intensive care unit. Photo taken by Dr. Daniel E. Keyler.

PLATE 8. W. H. Martin observing two timber rattlesnakes at a den in Maryland.
Photo taken on 9/3/1993 by William S. Brown.

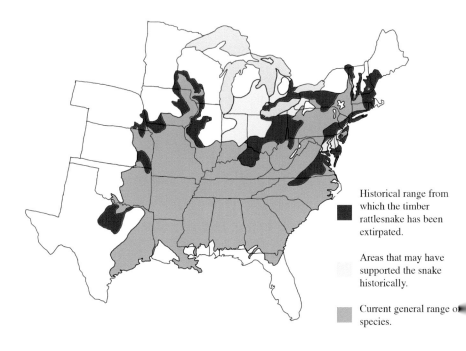

Historical range from which the timber rattlesnake has been extirpated.

Areas that may have supported the snake historically.

Current general range of species.

PLATE 9. The historical range of *Crotalus horridus*. Map drawn in 2005 by W. H. Martin.

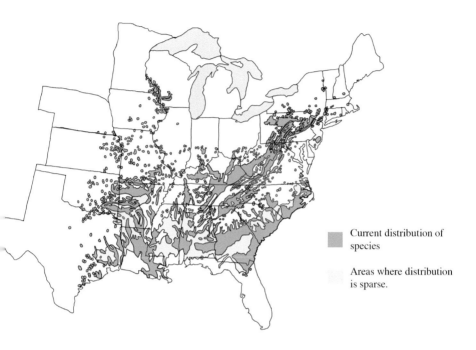

Current distribution of
species

Areas where distribution
is sparse.

PLATE 10. The current distribution of *Crotalus horridus*. Map drawn in 2005 by
W. H. Martin.

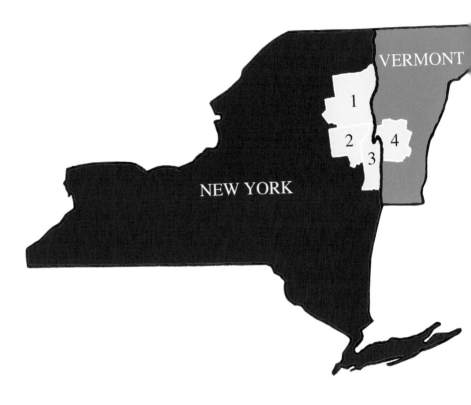

1. Essex County, New York

2. Warren County, New York

3. Washington County, New York

4. Rutland County, Vermont

PLATE 11. The four former timber rattlesnake bounty-hunting counties in Vermont and northeastern New York. Map drawn in 2006 by Jon Furman.

No one can say definitively how far back in time rattlesnakes go. There are rattlesnake fossils from Driftwood Creek in Hitchcock County, Nebraska, that are thought to be from the lower Pliocene or the upper Miocene era.[1] A rough estimate of the age of these fossils is somewhere from four to twelve million years. No one can pinpoint the origin of rattlesnakes either, but the prevailing opinion in the scientific community is that they originated somewhere in the northern mountains of what is now Mexico twelve to thirty million years ago and gradually fanned out from there. Some of these early rattlesnakes spread out in southerly directions, while others headed northward. There are rattlesnakes as far south as Argentina, and they populate large areas of South America. The bulk of the many species that are recognized today, however, occur in Mexico and the southwestern United States. At some point in time the forerunners of today's timber rattlesnakes probably migrated into what is now known as the Gulf Coast states region, and into the southeastern Atlantic Coastal Plain as well.[2]

The earliest rattlesnake fossils that can be labeled indisputably as those of timber rattlesnakes date back to what paleontologists refer to as the Irvingtonian II Land Mammal Age of the Pleistocene epoch, which translates out to about 400,000 to 900,000 years ago. Two locations in which these fossils have been discovered along with a wide variety of other herpetological fauna (herptofauna) are the Cumberland Cave site in Allegany County, Maryland, and the Hamilton Cave site in Pendleton County, West Virginia.[3] We simply don't know if timber rattlesnakes were in the Appalachian Mountains prior to 900,000 years ago, but it's pretty safe to infer, based on the fossil evidence from these caves, that they were in the central Appalachians around half a million years ago.

One may ask, how did the ancestors of the timber rattlesnakes that inhabit the mountains of Maryland, West Virginia, and other areas of the central Appalachians today survive the numerous ice ages of the Pleistocene epoch? The answer is that they probably didn't survive in that region during the coldest periods of the Pleistocene such as one that is believed to have occurred between roughly 23,000 and 25,000 years ago. Timber rattlesnakes require a mean July temperature of at least 65° Fahrenheit to survive, and summer temperatures this high are unlikely to have occurred in the central Appalachians 23,000 to 25,000 years ago. It was a brutally cold period of the Wisconsinan glacial age in which a mile-thick sheet of ice known as the Laurentide ice sheet reached as far south as southern Ohio, Indiana, and Illinois. In addition to the massive ice sheet, it is very likely that permafrost occurred along the higher crests of the Appalachian Mountains as far south as northern Georgia.[4]

Where and how, therefore, did the ancestors of today's central Appalachian timber rattlesnakes survive such conditions? Genetic evidence suggests that the major refuges in which they survived were in the lower Mississippi/Gulf Coast states region and in the southeastern Atlantic Coastal Plain.[5] W. H. Martin believes that a few isolated populations of timber rattlesnakes probably survived this cool-down in low-elevation valleys of the southern Appalachian Mountains in limestone caves, sinkholes, and fissures commonly referred to as karst.[6] Apparently, there are numerous areas of karst in the southern Appalachian valleys of northwestern Georgia, northeastern Alabama, and also in southeastern and south-central Tennessee. This karst extends northward into the Appalachians, but during the cool-down of 23,000 to 25,000 years ago, and in many previous severe cool-downs, it's very unlikely that any timber rattlesnakes survived except in the southernmost areas of karst. If they survived in those areas, the following pattern of survival probably took place repeatedly throughout the Pleistocene epoch: during every prolonged cold wave, enough timber rattlesnakes would survive in southern karst country to gradually penetrate into the central Appalachians as the climate warmed up afterward.[7]

There are no late Pleistocene sites of which I am aware in the central Appalachians that have radiocarbon dates in the neighborhood of 20,000 to 25,000 years ago as well as timber rattlesnake fossils. However, there are several sites in the region that have timber rattlesnake fossils and dates of about 11,000 years before present (ybp).[8] One of these sites is the New Paris 4 site in Bedford County, Pennsylvania. Another site is Clark's Cave in Bath County, Virginia. Assuming that remnant populations of timber rattlesnakes survived in karst regions of the southern Appalachians, as well as in the southeastern Atlantic Coastal Plain and in the lower Mississippi/Gulf Coast states region during the severe cool-down referred to above, they could easily have penetrated as far north as southern Pennsylvania by 11,000 years ago. They would have had 7,000 years since the glacial maximum of 18,000 ybp in which to accomplish this feat. If it started to warm up about 20,000 years ago, they would have had about 9,000 years in which to do so.[9]

It's difficult to say when timber rattlesnakes first appeared in the upper portions of the northeast. Over the last million years, numerous interglacial periods may have occurred in which the climate was warm enough for the species to penetrate not only into the central Appalachians but also into areas as far north as Vermont, upstate New York, and Canada. During a period about six to eight thousand years ago, the species probably maximized its range into southern Ontario, southwestern Quebec, north-central Vermont and New Hampshire, and southwestern Maine.[10] The earliest fossil evidence of timber rattlesnakes of which I am aware in the upper northeast occurs at a rock shelter site in Addison County, Vermont, and dates back to about 2,200 years ago.[11] Although this evidence may represent the oldest proof of timber rattlesnakes in the region, it does not disprove that timber rattlesnakes were as far north as southern Canada six to eight thousand years ago. The problem from a scientific point of view is that the places where timber rattlesnake fossils are most likely to exist in the upper portions of the northeast are deep down in talus slopes in dens that are impenetrable by man. Therefore, it's unlikely that anyone will ever be able to find any fossils in the area that could definitively prove the assumption that the snakes were as far north as southern Canada over six thousand years ago, but I think it's fair to assume that they were.

About five thousand years ago the climate began to cool down, and has been in a gradual and irregular cool-down ever since, until recently. Among other impacts, this long period of cooling very likely had the effect of shrinking the northern fringes of the timber rattlesnake's range. Martin believes that by the time the first European settlers arrived in the upper northeast, only isolated colonies, or populations, of these snakes remained clustered around the most favorable sites.[12] By some estimates the species was fairly widely dispersed in twenty-five or more sites when the first settlers arrived, in the early 1700s, in what has since become Vermont.[13] This makes three dens in western Rutland County an important natural resource, not only for Vermont, but also for New England at large, for two major reasons. They are the only known, active, and recovering dens in the entire state, and they also happen to hold the northernmost timber rattlesnakes remaining in New England.

Directly across the Vermont border in the eastern foothills of the Adirondack Mountains, a remarkably similar situation occurs. In Warren, Washington, and Essex counties in New York there are timber rattlesnake dens that, like the Rutland County dens, have probably been in use for the past six to eight thousand years or longer. Additionally, one of these dens has the distinction of being the northernmost stronghold of the species, not only in the state of New York but in northeastern North America as well. When taking a look at the current range of the timber rattlesnake (see pl. 10), it's apparent that two areas in the United States hold the northernmost timber rattlesnake populations, or groups of the snakes that hibernate in the same dens annually. One of these is the four-county area in the upper northeast referred to above. The other area is in the upper midwest in the states of Wisconsin and Minnesota. Timber rattlesnakes denning about ten miles below the forty-fifth degree of latitude in St. Croix County, Wisconsin, hold the distinction of being the northernmost timber rattlers in the country.[14] They are about twenty miles to the north of Minnesota's northernmost timber rattlers in Goodhue County, and about forty-five miles farther north than a population of timber rattlesnakes in Essex County, New York. Therefore, although the Essex County snakes are about as far north as this species occurs, they appear to rank as the third most northern population of timber rattlers in the United States.

One thing that's apparent when looking at a map of the current versus the historical range of the timber rattlesnake is that the species is far less widely dispersed in the upper northeast than it once was (see pl. 9). In New England, the snakes were extirpated, or hunted to extinction, in Maine by the 1860s, and in Rhode Island by the 1970s. As of the 1990s, the species was still surviving in New Hampshire but had dwindled to one confirmed colony in the southern part of the state. Massachusetts and Connecticut are faring slightly better but are down to a few dozen remaining colonies, primarily in their western mountains. There's a colony of timber rattlesnakes in eastern Massachusetts on a rugged promontory of land not far from the city of Boston. As previously mentioned, the species is confirmed as having survived in Vermont in Rutland County only.[15]

Not surprisingly, New York's timber rattlesnake numbers have also experienced a significant decline over the years. In the eastern foothills of the Adirondacks, W. S. Brown's field studies have revealed that denning populations are probably down by as much as 75% from what they were historically. From his long-term study of 139 dens, chiefly in the southern part of the state, Randy Stechert has determined that 76 dens have experienced depletion, in some cases to critical levels. Additionally, he has found that about a quarter of all the dens known historically in New York are now extinct. Overall he believes that the rattlesnake population of New York is only about 40% of what it was a hundred years ago.[16]

The obvious decline in timber rattlesnake populations in the northeast in the last few hundred years hasn't happened by chance, but rather because of man's encroachments, in one form or another, since the days of the earliest settlers. At least one of these encroachments, habitat destruction, hasn't shown any significant signs of letting up at the present time. If anything, it has intensified in the last fifty years. In the following two chapters I discuss the various encroachments that have reduced timber rattlesnake populations in the states of Vermont and New York, and the upper northeast in general, to a fraction of what they were historically. I have referred to this decline now simply to illustrate a point. Based on how long timber rattlesnakes have been in this portion of the country, and their tremendous decline in numbers over the last few centuries, it's reasonable to think of the remaining populations, which are widely scattered throughout the region, as survivors.

For one reason or another, man has been killing timber rattle-snakes directly and indirectly in the upper northeast for hundreds of years. Long before European settlers arrived in the region, various Indian tribes were known to kill these snakes for medicinal and other purposes. The Abenaki along with the Onondaga, one of the Five Nation tribes farther to the west, were known to use powdered timber rattlesnake flesh to cure fever. They were also known to use the smoke-dried gall of timber rattlesnakes to cure toothache.[1] Tooth health seems to have been an important matter to Native Americans in the upper northeast. In 1672 the English writer and naturalist John Josselyn reported, after living and traveling extensively in the area, that the Indians of New England would occasionally grab a timber rattlesnake by the neck and tail, tear off its skin with their teeth, and eat the flesh to refresh themselves while traveling. Whether they did this purely for sustenance is uncertain. They may also have done it to take on some of the rattlesnake's characteristics, and to protect their teeth.[2] It was a common practice for children of the Abenaki tribe to bite down on the entire length of a live, restrained timber rattlesnake from head to tail for the purpose of insuring sound teeth throughout the course of their lives. Other tribes in the northeast were known to use black snakes (black racers or black rat snakes, presumably) and in some cases green snakes (*Liochlorophis vernalis*) for this same purpose. The Mohegan were known to use black snakes, while the tribes of the Five Nations, the Iroquois, tended to use green snakes.[3]

It is the impression of Joe Bruchac, a well-known Abenaki story-teller from upstate New York, that the snakes used in the tooth strengthening practice or ritual were probably not killed as a rule, but they certainly might have been injured enough to cause lasting injuries and/or a shortened life. Joe related a remarkable story about a rattlesnake biting incident in which a large timber rattlesnake was killed.[4] The St. Francis Abenaki had a great chief from about 1750 to 1780 named Joseph-Louis Gill. Chief Gill was the child of white parents, who were captured and enslaved by the tribe, a fairly common practice at that time. We can only surmise how tough it was for him to grow up as a white child with blonde hair and blue eyes among a tribe well known for its toughness. It's likely that he was chided and tested frequently by his peers. As a young man around the age of twenty he was already well respected as a warrior when he did something that changed the tribe's opinion of him forever.

As the story goes, Joseph-Louis and a group of fellow warriors were about to leave for the Great Lakes region to fight the Miami, when one of his peers challenged his toughness. Apparently some timber rattlesnakes, one of which was exceptionally large, were sunning themselves on some nearby rocks. One way or another, his peers conveyed the idea to Joseph-Louis that he had to show the group of warriors that he was tough enough to bite down on the powerful, heavy-bodied snake from head to tail. After the other warrior pinned the big snake with a forked stick, Joseph-Louis grabbed it by its head and tail, held it up in the air, and broke its spine with one bite. He then proceeded to rip its body open with his teeth, tear out its beating heart, and swallow it. No one ever doubted the toughness of Joseph-Louis after that day. Furthermore, it showed his fellow warriors that he had what it took to become a chief.

Much has been written and documented about the Native American/rattlesnake relationship.[5] Certain tribes were known to wear rattles to fend off evil spirits. Various other tribes made their quivers out of rattlesnake skins. Many people are familiar with the famous Hopi snake dance in which a live, southwestern species of rattlesnake is held in the mouths of dancers. My interest for the purposes of this book centers more on the relationship that the Abenaki had with the timber rattlesnake specifically. The Abenaki were the primary Native Americans of Vermont and the three counties I refer to as the border rattler country of northeastern New York, i.e., Warren, Washington, and Essex counties. But no matter what tribe or what species of rattlesnake we're discussing, it's fairly safe to conclude that Native Americans, although they were known to kill rattlesnakes for a number of reasons, never seriously depleted any of the numerous species of these snakes the way the "white man" has. This stemmed from their respect for the snakes and, in some cases, their superstitious fear of retribution, if they were to kill rattlesnakes without a justifiable cause.

An example of Native American respect for and superstition about rattlesnakes was reported in a book by a man named Alexander Henry, a renowned fur trader of the Great Lakes region in the 1760s.[6] Henry was traveling with some Ojibwa when they came upon a rattlesnake, very likely a massasauga rattlesnake (Sistrurus catenatus).[7] He wanted to kill it, but the Ojibwa protested. Instead they surrounded it, called it grandfather, and blew smoke toward it. They then asked the snake to protect their families while they were gone, and set out in a canoe on Lake Huron. During their trip across the lake, a severe storm came up. The Ojibwa blamed it on Henry's having threatened the snake. In order to save their lives, they sacrificed two dogs by throwing them into the lake, offered some tobacco, and prayed to the rattlesnake god. One of the most interesting aspects of this account from my point of view is that the Abenaki, who had close cultural ties with the Ojibwa, may well have had the same super-stitious reverence toward rattlesnakes that they did.[8]

Where the Native Americans of New England may have had a respect bor-dering on reverence for the timber rattlesnake, this does not seem to have been the case with the European settlers, who have had a long history of reducing the snake's numbers ever since their arrival in the region. I am aware of four widespread encroachments that have been inflicted on the timber rattlesnakes of the upper northeast including New York, Vermont, and other parts of New England by people from largely European ancestry. None of these is more re-vealing than the organized, fear-driven timber rattlesnake hunts known to have taken place in Massachusetts in the seventeenth century.[9] They give us an insight into the way early European settlers and their descendants felt about rattlesnakes, and perhaps about snakes in general. It's anything but a feeling of respect bordering on reverence. Rather, it's almost the exact opposite, more akin to visceral fear and revulsion.

As Thomas Palmer pointed out in his book about the colony of timber rattlesnakes south of Boston, Massachusetts, there was a Puritan belief stem-ming from the Genesis story that snakes were analogous to Satan and there-fore a threat to, as well as an enemy of, mankind.[10] If the Puritans didn't have a strong religiously based fear of timber rattlesnakes, it's a little hard to imag-ine why they would take the time to hunt them, probably at their dens for the most part, and kill them off systematically until they became extinct in many locations. By all accounts, the Puritans led an extremely disciplined, hard-working life in which there would have been almost no time for anything other than attending to the daily necessities of their survival.

Were there fears other than those of a religious nature that motivated armed bands of Puritans to converge on timber rattlesnake dens to kill off as many of the creatures as possible? I believe there were. One of these would have stemmed from a perception that these snakes are much more aggressive and dangerous than they actually are. As aforementioned, this is a pretty common misconception. And no doubt the Puritans experienced the visceral, innate fear of snakes that so many humans seem to have, a fear that has been shown through experiments to occur in other primates as well. All in all, the Puritans had more than enough fear of timber rattlers to kill them wherever and whenever they could.

Although the religiously based fear of timber rattlesnakes may have ended in colonial times in Massachusetts, people's inherent fear of the snakes and the added fear stemming from misunderstanding didn't end then or there by any means. Gradually it spread throughout the upper northeast and elsewhere in the timber rattlesnakes' range, and it is still prevalent today. It will probably never cease altogether. Even in this more enlightened age, with laws in place to protect the snakes in many states, including Vermont and New York, many people will still kill one of these animals if it happens to come onto their property during its summer migration. Such a killing might be justified as an act of self-defense. No matter how much people love nature, or how enlightened they are in other ways, a timber rattlesnake seems a serious and frightening threat in the minds of people who kill one on their property. How little they know. I believe there are two ways in which the fear-driven killing of timber rattlesnakes could be greatly reduced. One way would be through strict enforcement of the laws already in place to protect the snakes; the other way would be through education. If people knew that they might be heavily fined if they killed a timber rattlesnake, or that the snakes play an important role in the food chains of the forests they inhabit and are only dangerous and aggressive when threatened themselves, far fewer would end up being clubbed to death, shot, or having their heads cut off by shovels.

In my opinion, there has been an overall failure in law enforcement. At the time of this writing, in only one case have charges been brought against an individual for illegally killing a timber rattlesnake in Vermont since the state listed the snake as an endangered species in 1987.[11] This case involved the killing of a timber rattler in the summer of 2003 in Rutland County. There have been almost no cases, probably fewer than eight, of individuals being charged for breaking the laws that protect timber rattlesnakes in the state of New York since the snakes were protected as a threatened species there in 1983. One of the cases that captured my attention involved a man who was bitten on the lip while showing off his snake-handling abilities to a group of friends while he was intoxicated. Apparently, he was charged while recovering from his envenomation in a hospital room.[12]

As for educating the public about basic timber rattlesnake biological realities, a lot of work needs to be done, especially in critical areas in the northeast and elsewhere where remaining populations of the snakes are in close proximity to civilization. In the border rattler country of northeastern New York and Vermont, timber rattlesnake seminars, well announced in advance and open to the public, are already being offered. As far as I know, W. S. Brown is the featured if not sole speaker at all of these functions. He's the perfect person for this job since he's a resident of the area, has spent his entire career studying timber rattlers in the area, and is one of the leading authorities on these snakes in the country. I have been to one of Brown's seminars on timber rattlesnakes in western Rutland County. Every seat in the house was taken, and a number of people were standing in the back of the room. Naturally, it's impossible for Brown in a half hour or so to delve into his long-lasting field study in any depth, or to cover everything interesting about the species he has come to know so well, but everybody in the audience that night in the spring of 2004 left the room with some good basic knowledge and, in many cases, less fear of the timber rattlesnake. They understood how minor a threat these snakes pose to human beings, and why the snakes need to be left alone and appreciated as an important part of the forests in which they live.

As successful as Brown's seminars are, only a fraction of the people who live or vacation in relatively close proximity to the timber rattlesnakes in the four counties on the Vermont/New York border attend them. Therefore, the majority of people in the area, who should know much more about the timber rattlesnake, remain uninformed about the species. How can this be rectified? One of the ways in which I can see the situation changing for the better is through a more intensified effort to educate the public. The press could provide a major cornerstone of this education, but has so far failed, in my opinion, to do so. The articles on timber rattlesnakes that appear in local newspapers and magazines from time to time tend to fall short in driving home three key points that could fundamentally change the public's fear of these snakes:

- they are only aggressive when threatened
- humans are rarely bitten by them, and those who are bitten almost never die as a result

- a timber rattlesnake that appears in somebody's back yard in the summer is very likely in the process of migrating through the area on its way to or from its den, and will likely soon go away.

There's no question that the more people know about these snakes, the less they'll fear them.

One of the things that has helped to minimize the fear-driven killing of timber rattlesnakes in northeastern New York was the establishment of W. S. Brown's "nuisance timber rattlesnake" program in the early 1980s. This program has been very successful at a number of campsites in the Adirondack State Park that are frequented by campers every summer. It has been successful because it has given campers an alternative to killing the timber rattlesnakes that occasionally appear on the islands, even if it hasn't lessened their fear of the snakes. As mentioned previously, when campers check in to these sites, they are informed by DEC campground rangers that they might encounter a timber rattlesnake during their stay, and they are told what to do if this should happen. It appears that the vast majority of campers have done exactly what they are supposed to do after encountering a timber rattler, which is to inform the rangers immediately. Over the last two decades, this has led to over a hundred so-called nuisance timber rattlesnakes being safely captured and relocated into the mountains, rather than possibly being wounded or killed by frightened campers.

Could Brown's "nuisance timber rattlesnake" program be expanded? Undoubtedly it could be. It wouldn't cost much for the few small towns in Vermont and northeastern New York that have populations of these snakes to train a few "nuisance rattlesnake" responders to capture, handle, and release these snakes safely. It would also be relatively inexpensive for these towns to mail some pertinent information regarding timber rattlesnakes to every resident, both year-round and seasonal. The information would need to include some key biological facts, give the legal status of the snakes, and give a number or two to call in the rare event that one of these animals might make an appearance on private property. In certain cases where timber rattlesnakes keep reappearing year after year on the same properties, it wouldn't be a bad idea to train the owners of these properties, as long as they were willing, to capture the snakes themselves. They would have to be shown how to pick up a timber rattler with snake tongs and put it into a garbage can or some other safe container until a responder could come and collect the snake.

At the Biology of the Rattlesnakes Symposium at Loma Linda University in the winter of 2005, it was interesting to see that there are other areas in the country where "nuisance rattlesnake" programs are being instituted and evaluated.[13] As man and the timber rattlesnake are placed in ever closer proximity in the northeast, it now appears from the success of Brown's program in the Adirondack State Park that other areas in Vermont and New York could effectively institute similar programs, and in so doing, save the fear-driven slaughter of many timber rattlesnakes over a period of time. Randy Stechert has successfully instituted a similar program in southern New York. As long as there is a local consensus that the timber rattlesnake is a worthwhile creature to protect, it would be foolish not to try these programs on a widespread basis, along with a concerted effort to educate the public about the species in sensitive locations wherever they occur.

Where fear-driven killing of timber rattlesnakes has been an ongoing encroachment from the days of the earliest settlers in the northeast, killing the snakes in order to produce rattlesnake oil was an encroachment that lasted for centuries but has virtually disappeared at the present time. Long before the European settlers arrived in New England, Native Americans in the area and elsewhere knew how to render a light, golden-colored oil from the fat deposits inside the body cavities of all rattlesnakes. John Josselyn gathered the impression from his encounters with New England Native Americans in the late seventeenth century that this oil was useful for treating aches, bruises, and even frozen limbs. How the substance was used depended largely on the particular tribe that was using it, as the following examples illustrate. The Ojibwa and the Fox used it as a muscle lubricant. It's possible that the Abenaki used it for this purpose as well because of their close cultural ties to the Ojibwa. The Cherokee apparently used it for the treatment of sore joints, while the Miami believed that the substance was effective in curing many different kinds of pain.[14]

As Laurence Klauber pointed out in his famous monograph on rattlesnakes, the medicinal use of snake oil was neither new nor confined to the Americas when the first settlers arrived on American shores. The colonists were apparently familiar with the beneficial use of viper oil, as it was a commonly used cure-all type of remedy in Europe long before they set sail for the New World. Once they arrived in the northeast and realized that various natives of the region were producing and using timber rattlesnake oil to treat some of the same conditions and ailments that people commonly treated with viper oil back home, they seem to have adopted the new viper oil as their primary remedy for a variety of ailments including pain and stiffness.[15]

When it came to producing rattlesnake oil, the early settlers probably didn't represent any more of a threat to the overall timber rattlesnake population of New England than did the various Native American tribes of the region, but once it was understood by the settlers sometime in the 1800s that timber rattlesnake oil had major commercial possibilities, the floodgates were open to hunting, killing, and depleting significant numbers of the snakes once again. It wouldn't surprise me if more timber rattlesnakes were killed to produce snake oil than were killed by roving bands of Puritans in the 1600s. Where the early round of killing was largely motivated by religion, the killing of the snakes for the purpose of making a commercially available oil was motivated by the desire to make money.

By the mid to late 1800s a number of small, local operations were bottling and selling snake oil liniment rendered from rattlesnake fat as a cure-all that could treat a wide range of conditions such as stiff joints, rheumatism, inflammations, pains of the back, and more. One company in the 1890s claimed that their snake oil would also cure headache, sore throat, neuralgia, sciatica, lumbago, calluses, corns, insect bites, frostbite, and deafness, among other things.[16] It's interesting to note that a number of the ailments that snake oil producers claimed their product would cure are common among people who work in the out-of-doors. Back pain, stiff joints, calluses, and farming, for example, go hand-in-hand. Apparently a sizable market developed for this product in rural America by the late 1800s and carried over well into the twentieth century. Snake oil liniment was widely sold and distributed by so-called medicine men in covered wagons in the formative days of the American west, but we know that the product was also being produced and sold on the Vermont/New Hampshire border, and presumably in other parts of the east during this period of time as well as earlier.[17]

By the time the snake oil industry was well established in the late 1800s, it's likely that the primary snakes used to produce the oil were eastern and western diamondback rattlesnakes (*Crotalus adamanteus* and *Crotalus atrox*, respectively). They are the number one and number two largest rattlesnake species in the United States, and therefore have much larger fat deposits than many of the other species of rattlesnakes, and the timber rattlesnake in particular.[18] In the early years of the commercial production of snake oil liniment in the mid-1800s, it's safe to assume that *Crotalus horridus*, the timber rattlesnake, was the primary, if not the only, rattlesnake used to produce oil. Why? The bulk of the American population still lived in the east, the oil was being produced in the east, and the timber rattlesnake was at that time, as it still is today, the most widely dispersed and common rattlesnake east of the Mississippi River.

How much of a decline in timber rattlesnake numbers did the early years of the snake oil industry create? We'll never know. There's no way even to make a rough estimate. It seems highly likely, however, that the burgeoning snake oil industry reduced the number of timber rattlesnakes in Vermont and New York and elsewhere in the east significantly, at least for a period of decades. One excellent piece of evidence from that era gives a lot of strength to this assumption. In his authoritative 1842 *Zoology of New York*, James Ellsworth DeKay refers to an article in the *Clarion* newspaper, published in Warren County, that throws some interesting light on the question of early snake oil production in that area.[19] According to the article, two snake hunters claimed to have killed 1,104 timber rattlesnakes in a recent three-day hunt on a prominent mountain range in the county for the purpose of producing snake oil. Assuming that two men were able to pull off such a feat, it's pretty obvious that they didn't kill this many timber rattlers to produce a little snake oil for their own use. Rather, they must have been involved in the commercial production of the oil, either as suppliers or producers, or possibly both.

I have made an all-out effort to determine the accuracy of this article because of what it tells us not only about the early days of commercial snake oil production in the northeast, but also about the historic number of timber rattlesnakes in northeastern New York State. I initially believed in its accuracy, and I still do, for two reasons primarily. First, we have to look at the number of snakes the two men claimed to have killed in their hunt. Liars or exaggerators could easily bark out a nice round number like eleven hundred when describing how many snakes they killed in a hunt. It would take a person or individuals more interested in accuracy and facts to come up with a figure such as 1,104.

Second, we need to consider the source of the reference to this hunt and the integrity of the *Clarion* itself. James Ellsworth DeKay was a highly esteemed American zoologist in his era. Based on the number of papers he published, as well as numerous other aspects of his career, it's fair to assume that he was meticulous with his references and therefore cited the *Clarion* article accurately. Among his other positions, Dr. DeKay served as both editor and librarian for the Lyceum of Natural History in New York City in the 1820s. When I learned that the *Clarion* was first published in 1841, I realized that Dr. DeKay had probably read about the hunt in an 1841 issue of the paper, but I didn't eliminate the possibility that he might have seen the article in an early 1842 issue of the paper.

Unfortunately, the New York State Library in Albany has only seven issues of the *Clarion* from the years 1841 and 1842 on microfilm, and none of these (I read them word for word) contained the article I so much wanted to find. All was not lost, however. While I was unable to find the article about the prolific snake hunt Dr. DeKay referred to, I was more than able to gather an impression of the quality and integrity of the *Clarion*. There is no question in my mind that it was an excellent newspaper. It was well written, interesting, and informative. It's hard for me to imagine that two local "boys" could have gone into the paper and pulled the wool over an editor's eyes about a seemingly unbelievable timber rattlesnake hunt without being totally believable themselves. They undoubtedly did kill what they thought to be 1,104 timber rattlesnakes, and there undoubtedly was a market as early as the 1840s in the border rattler country of New York and Vermont, and elsewhere in the east, for snake oil liniment derived from timber rattlesnakes.

One of the encroachments on the timber rattlesnake that has been around for years and represents a serious threat to populations of these snakes anywhere in their range, but especially to populations in the northeast that have been greatly depleted over a long period of time, is the capturing of live snakes for sale to collectors. Timber rattlers are sought-after by exotic pet fanciers because of their beautiful colors and patterns and adaptability to captivity. Where there's a market, there are dealers to supply the market. In 1991 timber rattlesnakes could be purchased from one Florida dealer for $40.00 each and for $45.00 from another.[20] The question is where did these snakes come from? If they came from New York where the snakes were listed as a threatened species on the state's endangered species list in 1983, or from Vermont, where they were listed as an endangered species in 1987, for instance, they were taken illegally. Additionally, it's illegal to sell timber rattlers from either state or transport them across state lines. Who knows where they came from?

During the last thirty to forty years, thousands of timber rattlesnakes have been captured in the northeast to sell to private collectors of one type or another. I am aware of two individuals who are thought to have taken five thousand or more timber rattlers between them for sale in the live animal trade in the region in the course of their careers. One of these men was the famous bounty hunter Art Moore, from Washington County, New York. As far as I know, Art did all of his live collecting of timber rattlesnakes within the confines of the law.[21] The other individual, whose name I choose to omit, may have stayed within the limits of the law in his early days of market hunting, but he clearly was not lawful and continued his activities illegally after timber rattlesnakes were widely protected throughout the northeast.[22] In 1993 this poacher served four months in a federal prison for illegally trafficking in timber rattlesnakes in the northeast. He was charged under the Lacey Act, which makes it illegal to transport any endangered species across state lines without a permit.[23] W. S. Brown, Len Jones, a U.S. Fish and Wildlife Service agent, and Randy Stechert wrote an article in which they detail the crime, how this man was set up, captured, and more.[24] This didn't stop his activities completely, however. As recently as 2001, he advertised the illegal sale of two hundred timber rattlesnakes through the Internet.[25] Laws in themselves won't stop poachers. They may deter the majority of them, but there will always be individuals who are willing to risk arrest (and even death in the case of African rhinoceros poachers) as long as the demand and reward are high enough. Ironically, the scarcer timber rattlers become, the more money they'll bring, and the greater the incentive to capture them in protected areas will become. It's a vicious cycle.

How much of an effect is the commercial collecting of timber rattlesnakes for sale to collectors having on the species in Vermont, New York, and other areas of the northeast at the beginning of the twenty-first century? Admittedly it probably isn't as serious as it was prior to the point in time when most of the states in the region listed their timber rattlesnakes as an endangered species. This is partly due to the fact that many of the markets for these snakes, such as roadside reptile farms, vanished long ago, but as W. S. Brown makes clear in his 1993 conservation guide, or "conguide" as he calls it, there is an active and sizable group of collectors world-wide who maintain a broad array of reptiles, including venomous species, in private collections.[26] It would be unrealistic to assume there isn't a current threat to timber rattlesnakes anywhere in their range, including the northeastern states where they are under protection by state and federal laws.

Unfortunately, we're at a point where timber rattlesnakes need all the help they can get in many areas, and taking one of these snakes out of its habitat for eventual sale to a collector is as serious as killing it. The removal of just a few sexually mature females from a small, remnant population of timber rattlesnakes could make the difference between the population surviving or becoming extinct. How do we as a society put an end to the poaching of these snakes? Although the laws that are in place to protect timber rattlesnakes won't stop this poaching, they can definitely help deter it, just as they can help deter the unnecessary killing of these snakes. The bottom line is enforcement. Every time the laws are enforced, and the cases involved are subsequently covered in the press, the less likely it becomes that people will break the laws in the first place. Another thing that can help is the development of monitors who regularly observe dens in critical areas. Many dens in the northeast already have dedicated individuals who have assumed the responsibility of protecting them from poachers and anyone else who might represent some form of threat to the snakes that overwinter in them. This is especially true in southern New York and in the border rattler country of Vermont and New York. In order to help protect this long maligned and persecuted animal in the northeast, more people should monitor more dens—people who not only care about the timber rattlesnake's survival but also understand how sensitive these snakes are to disturbance.

Of all the different ways in which man has encroached on and brought about the depletion of the timber rattlesnake in the northeast, none has been more widespread or caused the demise of more timber rattlers, directly and indirectly, than the destruction of their habitat. Nothing can hurt a species' survival more than taking away its habitat. For one reason or another, man has been destroying and adversely altering the habitat of the timber rattlesnake in the northeast for centuries. When the colonists first arrived in Massachusetts in the 1600s, they discovered vast, sprawling forests which they quickly began to clear. For approximately the next two hundred years, there was a tremendous demand throughout large sections of New England for lumber and open land. In those days almost everything was made of wood: the buildings, the furniture, the tools, and more. Once the forests were cleared, farms sprang up in the valleys, and sheep often grazed on the open hills and steep mountainsides much as they do in countries such as Ireland and Scotland today.

There is an interesting statistic about the landscape of Vermont in the late 1800s. By the time of the Civil War, about 85% of the land was open for sheep farming while 15% remained wooded. The woolen industry in New England needed wool, and Vermont apparently supplied a lot of it, but in so doing it severely deforested its landscape. Biologists know that this had a major impact on the wild turkey, for example, a bird that relies heavily on hardwoods for roosting and feeding. By 1870, wild turkeys were extinct in Vermont. There's absolutely no doubt that this deforestation had an impact on the state's timber rattlesnakes as well. Dens that had previously been hidden safely away in wooded, mountainous habitats were suddenly exposed. Whenever a den was discovered, the colony of rattlesnakes that overwintered in it was probably ex- terminated within a few years owing primarily to the fear that the settlers had of the snakes. Killing off the snakes to produce rattlesnake oil may have fac- tored in as well. Zadac Thompson made a reference in a history of Vermont, which he wrote in 1842, to the effect that there had been considerable numbers of timber rattlesnakes in the state in the past but that they had been greatly re- duced by the inroads of man's activities. He neglected to mention what those activities were.[27]

By the late 1800s New England and the entire northeast began to shift from an agricultural base and enter into the emerging Industrial Revolution. During this era many people gave up farming to work in factories and mills in the larger towns and cities. Countless small farms were sold off and the extensive forests of colonial times began to return throughout the region. One of the upshots of this was that for the next seventy years or longer timber rattlesnakes had a little reprieve in their remaining strongholds. This is not to say that fear-driven killing of the snakes decreased in any way, but their habitat was pretty much left alone until the second half of the twentieth century. In the early 1950s a new type of habitat destruction began to emerge, with the result that thou- sands of timber rattlesnakes would suffer serious consequences in many areas.

After World War II many people in urban parts of the northeast became in- volved in what has often been referred to as a return-to-nature movement. One of the major manifestations of this movement was the building of second homes and summer camps in pristine mountainous areas that had had popu- lations of timber rattlesnakes for thousands of years.[28] Another manifestation was that people began to hunt and fish and recreate more in these areas. The huge popularity of jeeps after the war subtly factored into this movement by enabling people to get back into areas that had previously been left alone.

At first the effect on the timber rattlesnake was somewhat minimal, but this pattern has changed, as the movement has continued to grow to an extent that was probably never dreamed possible fifty years ago. Long gone are the days of building primarily around mountain lakes for the purpose of recreation and retreat. In the 1960s and 1970s housing developments began to appear in mountainous areas without lakes. For the most part they blended in pretty well, disrupting only minimally the immediate environments around them. Nowadays huge areas on the sides of mountains are clear-cut and bulldozed to make room for sprawling homes and condominiums.

Along with this massive destruction of timber rattlesnake habitat in many cases, motorized off-road machines (dirt bikes and four-wheelers) scream through wooded areas, going almost anywhere their drivers choose to take them. The upshot of all this is that a lot of timber rattlesnakes have already been killed off, and many colonies of these snakes will be in jeopardy in the future despite being widely protected throughout the northeast. Man has clearly moved in too close for comfort in many areas, not only through development but also through various forms of recreation. Consequently, a lot of pristine timber rattlesnake habitat has been destroyed or adversely affected, and the ongoing destruction of this habitat continues to be the major threat to the timber rattlesnake's survival. Nowhere in the northeast is this more evident than it currently is in southern New York State.

CHAPTER 6 : BOUNTY HUNTING

Ironically, although habitat destruction has been and will continue to be the major human encroachment to threaten timber rattlesnakes in the northeast as a whole, it doesn't appear to have been a significant source of depletion for the border rattlesnakes of Vermont and northeastern New York. For that matter, I doubt if any of the encroachments I have detailed thus far, with the possible exception of killing the snakes to produce rattlesnake oil, had much to do with the depletion in this area. This makes the timber rattlesnakes of Rutland County and the eastern foothills of the Adirondacks unique in the northeast. The rugged mountainsides where they have lived for thousands of years have been impenetrable to all but the hardiest woodsmen and, therefore, have served to isolate the snakes from all human encroachments except for one: bounty hunting.

The most aggressive and sustained encroachment in the history of man in the upper northeast, or anywhere else in the timber rattlesnake's entire range for that matter, was unquestionably the bounty hunting period that lasted from the late 1800s until the early 1970s in the four-county area I describe as the border rattler country of Vermont and New York.[1] During this era, thousands of timber rattlesnakes were killed by a relatively small but highly successful and independent group of individuals. This period of history is of particular interest to me because of the impact it had on the rattlesnakes of Rutland County, Vermont, and Warren, Washington, and Essex counties in New York. If bounty hunting had lasted past 1973 in New York, the snakes might well have been extirpated or be on the verge of extinction in the eastern foothills of the Adirondacks, despite the rugged habitat in which their dens are typically situated. By the 1950s several of the largest dens in the area had already been hunted to extinction. If the snake hunting had lasted past 1971 in Vermont, I believe that the snakes would be barely hanging on today in the state's three active dens in the western part of Rutland County.

New York never had a statewide bounty on timber rattlesnakes but probably could have had one. To my knowledge, only three counties in the northeastern section of the state ever paid money for dead timber rattlers. The species can be found in other parts of the state such as the Southern Tier, or the south-central and southwestern uplands along the Pennsylvania border, the foothills of the Catskills, and the Hudson Highlands, but people in these areas never seemed to be particularly concerned about a rattlesnake problem. At some point in the early to mid-1800s the inhabitants of the town of Stillwater in Saratoga County, New York, may have sensed that they had a problem with timber rattlesnakes. Joe Bruchac told me about a hermit who supposedly lived on Snake Hill in Stillwater and hunted timber rattlesnakes there for a bounty back in the middle part of the nineteenth century.[2] Although I have been unable to confirm such a bounty through written records, there may well have been a local timber rattlesnake bounty in the town of Stillwater at that time. This could at least partly explain the extirpation of the county's only rattlesnake den by sometime in the mid-1800s.[3] The den was located in Stillwater, on Snake Hill, above the eastern shoreline of Saratoga Lake.

Through its County Board of Supervisors, Saratoga County enacted a bounty on timber rattlesnakes that lasted for only three years, from 1948 through 1950. This is absolutely remarkable, since timber rattlesnakes were only ever known to exist in one location in the county, at the long ago extirpated Snake Hill den. Perhaps the Board of Supervisors assumed that rattlesnakes were still alive and well in the county in 1948, since three of its neighboring counties had bounties on the snakes at that time. Since there are no longer any members of the County Board of Supervisors alive from that era, it's impossible to say definitively what motivated the board to enact the bounty, or, for that matter, what motivated them to discontinue it after such a short period of time.

For the purposes of this book I am more concerned with three of Saratoga County's neighbors to the north and east. Apparently there were enough timber rattlers in concentrated areas of Warren, Washington, and Essex counties in the latter part of the nineteenth century to convince local residents and politicians that they had a serious rattlesnake problem. These people may well have believed that the snakes would, among other things, adversely affect the important summer tourist trade on the lakes in the region. Essex was the first of the three counties to take action by initiating a bounty on the snakes in 1892. Washington County followed suit in 1894, and Warren County followed in 1896.[4] We know that timber rattlesnakes have survived in isolated populations in the mountains of this vast area. The question is, how large were these populations at the time of the bounty enactments? As previously mentioned, the *Clarion* newspaper article in 1841 or 1842, which referred to two men killing 1,104 timber rattlesnakes in a three-day hunt on a mountain range in Warren County, throws some light on rattlesnake oil production in the northeast in the middle part of the nineteenth century. At the same time it gives us an idea of the overall population of timber rattlers in Warren County at that time. If the two snake hunters killed even half the number of rattlesnakes they claim they did, the population was an abundant one. Assuming a mountain range so abundant with timber rattlesnakes midway through the 1800s, it's fairly safe to project a large local population of these snakes on that range, and possibly at other locations in the county, in the late nineteenth century when the county's long-lasting timber rattlesnake bounty hunting period began.

One of the interesting facts about the timber rattlesnake bounty in Warren County is the way it gradually increased monetarily over the years. In Resolution 21, adopted unanimously by the County Board of Supervisors in 1896, a person could receive twenty-five cents for a dead timber rattlesnake with one rattle or button, and one dollar for any snakes with two or more rattles. What they meant by this was that they would pay twenty-five cents for a neonate with a button rattle and a dollar for an older snake with two or more segments in its rattle.[5] In 1924 the bounty was raised to a dollar and a half per snake. (By that time in the evolution of the county's timber rattlesnake bounty, the board was no longer concerned about the number of segments a rattlesnake had in its rattle.) In 1928 the bounty was raised to three dollars. It dropped back to a dollar and a half in 1932. In 1936 it jumped up to two dollars and fifty cents per snake, where it remained until 1948 when it went back up to three dollars. In 1951 it was raised to five dollars per rattler and stayed there through 1973.[6]

Although New York State discontinued its bounties on all species statewide in Section 206 of its Conservation Law in 1971, Warren County, via Resolution 77, adopted by its County Board of Supervisors in 1971, and Resolution 107, adopted in 1972, continued to make bounty payments on timber rattlesnakes through 1973.[7] These resolutions indicate that the powers-that-be in Warren County in the early 1970s still had some serious concerns about their timber rattlesnake population. Resolution 107 is very explicit about the board's two

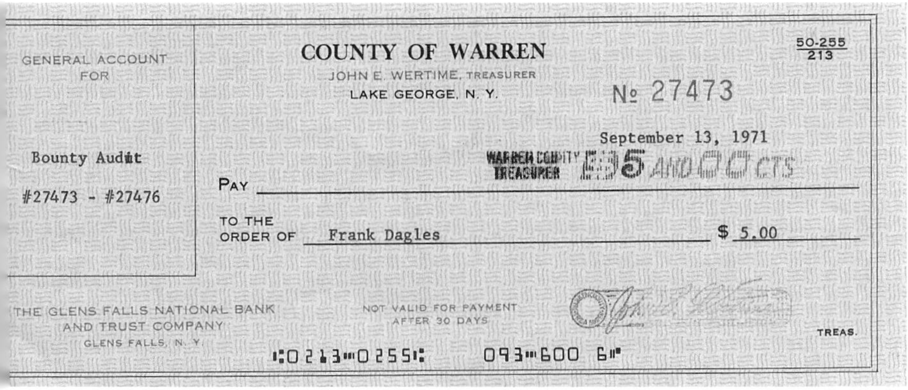

GENERAL ACCOUNT
FOR

Bounty Audit

#27473 - #27476

COUNTY OF WARREN
JOHN E. WERTIME, TREASURER
LAKE GEORGE, N. Y.

50-255
213

N⁰ 27473

September 13, 1971

PAY

WARREN COUNTY
TREASURER

TO THE
ORDER OF   Frank Dagles $ 5.00

THE GLENS FALLS NATIONAL BANK
AND TRUST COMPANY
GLENS FALLS, N. Y.

NOT VALID FOR PAYMENT
AFTER 30 DAYS

TREAS.

⑈0 2⑈3⑈0 2 5 5⑈⑈   0 9 3 ⑈ 6 0 0   6⑈⑈

FIGURE 2. A September 1971 uncashed bounty check for one timber rattlesnake, made out to Frank Dagles of Warren County, New York.

principal concerns. The first had to do with health, arising from the fear that the species could in some way pose a health threat, or hazard, as they worded it. Their other concern was economic. In particular, they were afraid that the snakes would lower real estate values and discourage summer tourism. Art Moore was probably correct when he told me in my first interview with him at his home in the summer of 2002 that the last thing a summer tourist in his area wanted to see was a timber rattlesnake, and that this fear was what had started the rattlesnake bounties in Warren, Washington, and Essex counties in the first place. Art had a way of hitting the nail on the head.

One of the most curious aspects of the long period of timber rattlesnake bounty hunting in Warren County, from 1896 through 1973, is that although numerous resolutions were adopted by the County Board of Supervisors to modify and prolong the bounty on the snakes, no resolution was taken to discontinue the bounty. Resolutions 77 (1971) and 107 (1972) extending the bounty were two of the last three resolutions dealing with the rattlesnake bounty ever enacted in the county (see Appendix A). Since they are completely open ended, it's a little puzzling why the county stopped paying a bounty on timber rattlers after 1973. We know through W. S. Brown's examination of bounty payment records early in his career, when many records existed that are now no longer available, that 1973 was the last year in which Warren County made bounty payments on timber rattlesnakes.[8] It's my assumption that, for one reason or another, the powers-that-be in the county decided in 1974 to abide by Section 206 of the state's 1971 Conservation Law, and discontinue its extended bounty on rattlesnakes just the way it had discontinued its bounty on "wolves, coyotes, coy dogs and bobcats" three years earlier.

Did bounty hunting significantly deplete the timber rattlesnake population of Warren County? It could be argued that the county's overall timber rattlesnake population was large enough not to be seriously affected by the bounty hunting period. For two reasons, however, this argument won't hold water. First, we have to consider Art Moore, who, as previously mentioned, was probably the most prolific timber rattlesnake bounty hunter in New York State history. Other bounty hunters in New York killed a lot of timber rattlesnakes in their careers, such as the legendary Will Clark, Jr., of Warren County (see fig. 3), but Moore was the sole prolific killer of these snakes in what could be thought of as the modern era of timber rattlesnake bounty hunting in the state, the 1950s through the early 1970s. Of the fifteen to eighteen thousand timber rattlers he is unofficially credited with killing in his long career in the state's three bounty counties, we know that he killed many of those snakes in Warren County.[9]

Because of incomplete records, we'll never know the exact number of rattlesnakes he killed there, and it really doesn't matter. What we know is that the current estimated timber rattlesnake population in the entire state of New York at the beginning of the twenty-first century is only between eight and ten thousand snakes.[10] Since Art Moore killed thousands of timber rattlers in Warren County in the period of time from the early 1950s through the early 1970s, it's safe to say that bounty hunting significantly reduced the number of timber rattlesnakes in the county during those years. We know, for instance, that Moore killed 2,459 snakes in the last seven years alone of the rattlesnake bounty hunting period in the county.[11]

There is another way to visualize the impact of bounty hunting on Warren County's timber rattlesnake population. It's a subtle detail that has emerged from W. S. Brown's extensive and ongoing field study in the county. When Brown began his research in 1978, very large timber rattlers were almost non-existent. A dozen or so years later, in the early 1990s, he began to see and capture some large snakes on occasion. By the end of the 1990s, he was capturing and measuring some of the longest snakes of his career, and large males, which grow bigger than females, were becoming an increasing proportion of his total catch.[12] What this indicates is that bounty hunting, especially in the 1950s, '60s, and '70s, had taken a major toll on big male rattlers in the various local populations of snakes he's been studying for years.

FIGURE 3. Well-known rattlesnake bounty hunter Will Clark, Jr.,
in Warren County, New York. Photo taken in 1948 by Richard K. Dean.

Brown's study, among many other findings, has established that timber rattlesnakes are capable of living twenty to thirty years in the wild. It has also shown that a male of the species needs over two decades to mature to its maximum size. A male born between 1974 (a year after the timber rattlesnake bounty extension was discontinued in Warren County) and 1984 would reach full maturity between 1994 and 2004. It all makes sense. In the first few years of the new millennium Brown has been recapturing some of his marked adult males that are over twenty years old, and a few that he can document to be over thirty years old. Snakes like these weren't around when he began his career in the late 1970s. It's obvious why. Art Moore, and approximately two to three dozen lesser bounty hunters, had killed them off between 1951 and 1973 when the bounty in Warren County was at five dollars per snake.[13] Because of their size, these snakes, the easiest to see and kill, disappeared rapidly under heavy bounty hunting pressure.

Through the mid-nineteenth century hunt in which 1,104 rattlesnakes were reported killed by two men in three days, we have a way of projecting the overall population of timber rattlesnakes at the end of that century in Warren County. It would be helpful if we had some comparable hunt or hunts in the middle to late 1800s in adjoining Washington County to help us envision the overall population there when the County Board of Supervisors enacted its first rattlesnake bounty in 1894. My guess is that it was sizable. I base this partly on the fact that Washington County enacted its bounty on timber rattlesnakes two years before Warren County did, so presumably the county supervisors were perceiving, through numerous rattlesnake sightings and reports, a serious rattlesnake problem. I don't believe, however, that Washington County had nearly the overall number of timber rattlesnakes that Warren County had at that time, for the same reason that the two counties have disproportionate overall populations of timber rattlesnakes today. Washington County has a smaller number of known dens than Warren County has.

Whereas Warren County enacted numerous monetary changes to its timber rattlesnake bounty over the seventy-six years in which it was in effect, this was not the case with Washington County. During the seventy-eight years in which its bounty was in effect, only a handful of changes were made affecting how much a hunter could collect per snake. In what was termed Act No. 1, a law adopted by the County Board of Supervisors in 1894, a person could receive fifty cents for any rattlesnake killed within the county. Nothing was contained in the resolution about the size of the rattle. The only reference to rattles was that they had to be delivered to the supervisor of the town in which the snakes were killed, and thereupon destroyed by him. For over fifty years the bounty remained at fifty cents. Then in 1948, in coordination with the County Boards of Supervisors from Warren and Saratoga Counties, the Washington County Board of Supervisors raised its timber rattlesnake bounty to three dollars per snake. It remained at that level until 1956 when it went up to five dollars per snake, and remained at five dollars through 1973.[14]

Washington and Warren counties have some remarkable parallels in the wording of some of their final rattlesnake resolutions. In Resolution 184 of 1971 (see Appendix B), the Washington County Board of Supervisors made a request to the New York State Legislature to introduce a bill that would override Section 206 of the state's Conservation Law, the law that had put an end to all bounties statewide as of July first of that year. To support their request, the supervisors cited exactly the same issues that the Warren County Board of Supervisors would cite in its Resolution 107 adopted the following year.[15] They were clearly concerned with the negative impact the snakes could have on real estate values, particularly if a child were to be bitten and killed by one of the snakes. They depicted the snakes as presenting a health hazard in general, and made the point that the danger of someone getting bitten increased during the summer months owing to an influx of tourists in the area. Their major concern, although they were careful not to make it too obvious in their wording, was that rattlesnakes could hurt the county financially. This same concern reappeared in the spring of 1972 in Resolution 73 (see Appendix C), which also overrode Section 206 of the state's Conservation Law of the previous year and reinstated the bounty on rattlesnakes in three of the county's towns.[16]

Another striking similarity between Washington and Warren counties at the end of their long-lasting rattlesnake bounty hunting periods is the way they suddenly stopped making payments on the snakes after 1973. W. S. Brown made this discovery about both counties discontinuing their timber rattlesnake bounties during his examination of bounty payment records in the late 1970s.[17] Even though there never was a resolution to discontinue the bounty on rattlesnakes in Washington County, I believe, as I said of Warren County, that the authorities must have decided in 1974 to abide by Section 206 of the state's Conservation Law of 1971, and discontinue bounty payments altogether.

The records are incomplete here, too, about whether bounty hunting significantly depleted Washington County's overall timber rattlesnake population. My answer would be that it did. We have some excellent clues. Again, we have to look at the prolific bounty hunting career of Art Moore before considering anything else. Although the bounty payment records were incomplete when W. S. Brown examined them in the 1970s (there were no records for the years 1964 through 1967 whatsoever), it was easy to see from them that Moore had made the bulk of the 5,611 rattlesnake bounty claims that Washington County paid during the years 1953 through 1963, and 1968 through 1973.

It's not important that we know Moore's exact career numbers in the county. Bearing in mind Randy Stechert's current estimate of seven to ten thousand timber rattlesnakes remaining in the entire state of New York, and Moore's share of approximately 80% of the 5,611 bounties claimed (and possibly thousands more during the four years 1964 through 1967), it's entirely possible that Moore killed almost as many timber rattlers in Washington County during his career as exist in the entire state of New York today. It seems fair to assert, therefore, that Moore alone, or bounty hunting as a whole, overwhelmingly reduced the timber rattlesnake population in Washington County in the bounty hunting years.

Art Moore was certainly not the only timber rattlesnake bounty hunter who hunted in Washington County, or in Warren County for that matter, but since he killed far more timber rattlesnakes in those counties than anyone else during the last three decades in which bounties were offered, it's easy to overlook the roughly twenty to thirty other individuals who hunted during that period as well. These men were much more casual as rattlesnake bounty hunters than Moore was. Not one of them seems to have matched Moore by as much as 10% in bounty claims from the 1950s through the early 1970s. Yet we need to realize that these hunters collectively killed hundreds of snakes in Washington and Warren counties, and in so doing created a serious reduction in both counties' timber rattlesnake populations.

Of the three important New York timber rattlesnake bounty counties, we know that Essex had the lowest overall population of snakes historically, owing to the fact that the county has only one known denning area. This is the den I referred to previously as the northernmost stronghold of these snakes not only in New York State but also in northeastern North America. If Essex County had had a large population of timber rattlers at the turn of the nineteenth century, I believe we would see an indication of this in the evolution of the county's bounty hunting resolutions during the next seventy some years. It would have looked more like Warren County's evolution over the same period of time.

As discussed, Warren County made numerous monetary increments to its rattlesnake bounty during the 1920s, '30s, and '40s. But this didn't happen in Essex County, whose Board of Supervisors apparently never felt that the county's rattlesnake population was getting out of hand during those years, and therefore never made any changes to the one dollar per snake bounty it enacted in 1892, until it increased it to five dollars in 1956.[18] Did the members of the board suddenly think in 1956 that the county was overrun with timber rattlesnakes to the point where it needed to enact such a significant increase to the bounty it had been offering on the snakes for over sixty years? I don't think so. I believe that the driving force behind the decision to increase the rattlesnake bounty in 1956 was the board's desire and willingness to fall in line with Warren and Washington counties' five-dollar bounties.

One of the revealing facts about the evolution of the rattlesnake bounty in Essex County is that the Board of Supervisors never bothered to enact any resolutions that would override Section 206 of the state's Conservation Law of 1971 so as to extend the county's bounty on rattlesnakes only. The last resolutions enacted in the county that had anything to do with timber rattlesnakes were written in 1969.[19] From this we can pretty safely infer that there wasn't an abundance of timber rattlers, and thus a rattlesnake problem, in the county at that time. It's a little difficult to say whether bounty hunting seriously depleted Essex County's timber rattlesnake population. To my knowledge there are no bounty payment records or other sources of information that would give us an indication of how many timber rattlers were killed in the county during the bounty hunting years. My educated guess is that bounty hunting in Essex County reduced the county's overall population of timber rattlers little by little over the years, but that there was never a time when these snakes were seriously depleted by bounty hunting. Something that may have seriously depleted the snakes was a sudden, wind-driven forest fire in the 1960s. It was Art Moore's belief that a lot of snakes were killed in their summer habitat when this fire broke out, and there was nowhere for many of them to hide. Moore said he didn't hunt very often in Essex County, and he only knew of one other bounty hunter who hunted there. He also claimed that he stopped bounty hunting altogether in the county after the fire, as did the other man, owing to the fact that the snakes seemed to have mysteriously vanished.[20]

When the Vermont Legislature enacted a bounty on timber rattlesnakes in 1895, the bounty was statewide.[21] It probably didn't need to be, though, since the only real concentration of rattlesnakes remaining in the state at that time was in the three dens previously referred to in Rutland County. A few other areas in the state had populations of timber rattlers, such as Skitchewaug Mountain in Windsor County, but they were tiny remnant populations that eventually disappeared altogether through deliberate, fear-driven killing rather than bounty hunting. As far as I know, Rutland County is the only county in Vermont in which bounties were ever paid for dead timber rattlers.

One of the most striking and revealing aspects of the long timber rattlesnake bounty hunting period in Vermont, other than the fact that it basically applied to one county, is the fact that the Vermont Legislature never deemed it necessary to increase its bounty on timber rattlesnakes the way Warren, Washington, and Essex counties did in New York. The bounty on timber rattlesnakes in Vermont began at one dollar per snake in 1895 and remained at one dollar per snake until the state discontinued its bounties on all animals in 1971. This suggests that the representatives in the three small towns in which rattlesnake bounties were paid in Rutland County never sensed that their towns had a large overall population of timber rattlesnakes and/or a serious rattlesnake problem.

Vermont lost a lot of timber rattlesnakes to bounty hunters, but it's difficult to determine just how much of a depletion to the state's overall timber rattlesnake population was brought about by these individuals. Of the three rattlesnake dens in Rutland County, the smallest one is so isolated and unknown in general that it probably had very little bounty hunting pressure at all.[22] The other two dens are much larger, and I suspect that bounty hunting seriously reduced the timber rattlesnake population at the largest of these dens since it happens to be on the former property of Vermont's best known bounty hunter, Bill Galick. Starting in 1946, Galick and his brother, Ed, logged in far more rattlesnakes in Vermont than anyone else in the previous fifty years, and as far as I know through many interviews and conversations with Bill, they never hunted the snakes anywhere except on their own property.[23]

Despite the fact that the second largest den in Rutland County was more accessible to bounty hunters than Galick's den was, it was probably much less affected by bounty hunting because no prolific hunters of the caliber of Art Moore or Bill Galick ever seemed to develop at this site. I ascertained this by examining old bounty payment records from the town in which the den is located in western Rutland County, and by interviewing former bounty hunter Lester Reed, who had hunted the den exclusively.[24] Between 1899 and 1946 the records

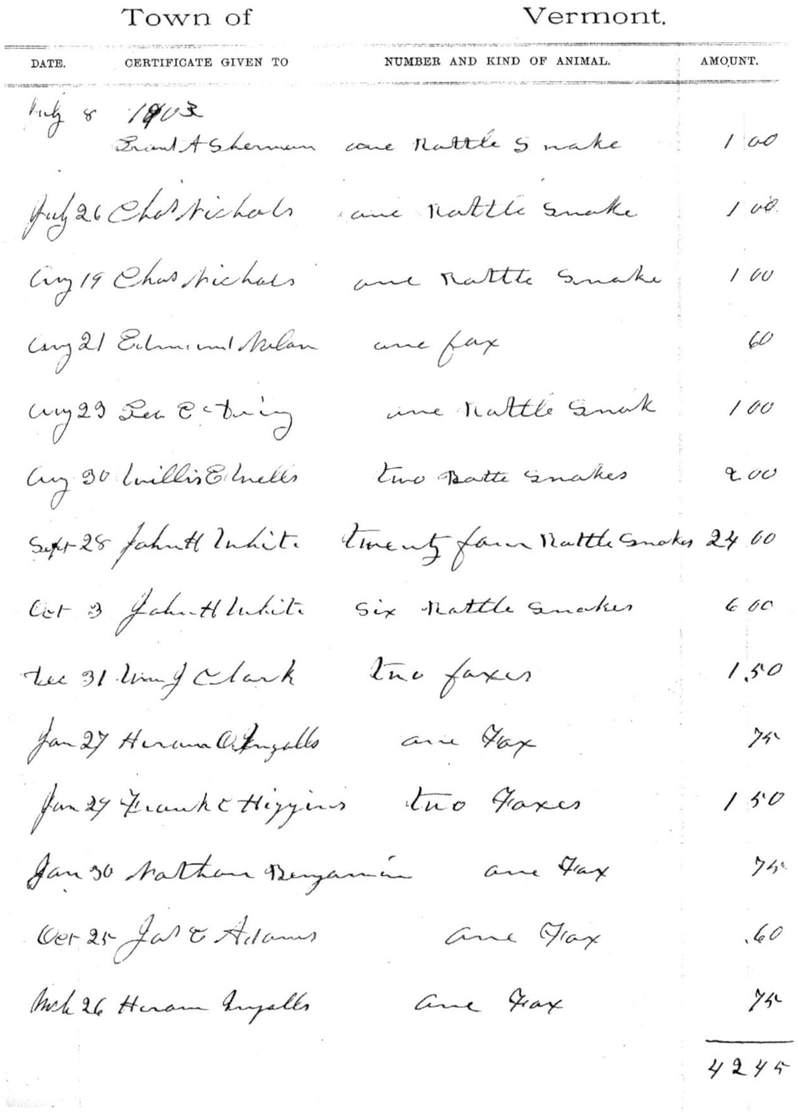

Town of   Vermont.

DATE.	CERTIFICATE GIVEN TO	NUMBER AND KIND OF ANIMAL.	AMOUNT.
July 8	1903		
	Frank A Sherman	one Rattle Snake	1 00
July 26	Chas Nichols	one Rattle Snake	1 00
Aug 19	Chas Nichols	one Rattle Snake	1 00
Aug 21	Edmund Nolan	one fox	60
Aug 23	Leo C Twing	one Rattle Snak	1 00
Aug 30	Willis E Snells	two Ratte Snakes	2 00
Sept 28	John H White	twenty four Rattle Snakes	24 00
Oct 3	John H White	six Rattle Snakes	6 00
Dec 31	Wm J Clark	two faxes	1 50
Jan 27	Hiram A Ingalls	one Fox	75
Jan 29	Frank C Higgins	two Foxes	1 50
Jan 30	Nathan Benjamin	one Fax	75
Oct 25	Jas E Adams	one Fox	.60
Mch 26	Hiram Ingalls	one Fox	75
			42 45

FIGURE 4. A 1903 bounty payment record from western Rutland County, Vermont.

reveal that a few snakes were taken, presumably from this den, by a small number of casual hunters each year. According to Lester Reed, this pattern of minimal but steady depletion continued there from the late 1940s until the end of the bounty hunting period in 1971. My assessment of Vermont's timber rattlesnake bounty hunting period is that it did seriously deplete the state's rattlesnake population. Whether it brought about a 25% depletion or a 75% depletion, its overall impact was substantial. When there are only three active timber rattlesnake dens left in a state, killing any of the snakes from those dens, particularly pregnant females, is seriously detrimental.

In researching the history of timber rattlesnake bounty hunting in Vermont and New York, it became apparent on a number of occasions that the bounty systems in both states were far from perfect and therefore quite vulnerable to fraud. One of the most common frauds was what I refer to as border-jumping. This was the practice of Vermont bounty hunters taking the tails and rattles of snakes they had killed in Vermont across the border into New York, claiming that the snakes were killed in New York, and collecting the higher New York bounties. Rutland County bounty hunters in the 1950s knew full well that by taking their rattlesnake remains to a town clerk in Warren County, starting in 1951, they could get five dollars per snake instead of one dollar back home. They also knew they could get five dollars per snake in Washington or Essex County, starting in 1956.[25] How did they accomplish this fraud? The trick was to have a friend in New York who was willing to go to one of the town clerks in either county with a bag full of tails and rattles from timber rattlesnakes that had been killed in Vermont by another individual, and claim that he, the New Yorker, had killed the snakes in that county. After filling in the necessary forms, the New Yorker would eventually receive a check from the County Treasurer of the county in which he had made the claim. At that point he would cash the check and give his Vermont friend the proceeds. It was all quite illegal, but I've been told by a former Vermont bounty hunter, who routinely border-jumped rattlesnakes in his "day," that the New York clerks were very lenient.[26]

Another timber rattlesnake bounty hunting fraud similar to border-jumping took place routinely in New York. It also involved misrepresentation of fact in order to collect a rattlesnake bounty, but Vermonters probably had very little if anything to do with it. This fraud mainly involved New York bounty hunters. Starting in 1969 in Warren County, and 1970 in Washington County, the only people eligible to collect a bounty on timber rattlesnakes in either county were bona fide residents of those counties. This made it a little tricky for Art Moore and others to kill rattlesnakes outside their home counties and collect the bounty within them with no questions asked or suspicions raised.

In one of my early interviews with Moore in the summer of 2002, he explained the residency requirements referred to above. Knowing that he was a lifelong resident of Washington County who hunted timber rattlesnakes exhaustively in Warren County through 1973, I asked him how he dealt with the residency issue when he hunted and reported snakes in Warren County. He said it was very simple. Whenever he hunted rattlesnakes in Warren County in the final years of its bounty on timber rattlesnakes, all the snakes he killed there were turned in by any one of several friends he normally hunted with who were residents of the county. As Art explained this to me, he emphasized the fact that the town clerks in New York were easy to deal with when it came to making timber rattlesnake bounty claims. It was his impression that they weren't particularly fond of the snakes, and that they weren't about to blow the whistle on a timber rattlesnake bounty hunter whose claim, or multiple claims over a period of time, may have seemed a little questionable to them.

A third way in which people tried to defraud the timber rattlesnake bounty hunting system in northeastern New York was very similar to border-jumping in that it involved taking timber rattlesnakes that had been killed in another state or another part of New York, to a town clerk in one of New York's timber rattlesnake bounty hunting counties, and claiming that the snakes were killed in that county in an attempt to receive the rather lucrative payment of five dollars per snake. The difference in this fraud was that it didn't involve Vermont bounty hunters working with New York accomplices. It also tended to involve very sizable claims, and in some cases it didn't involve bounty hunters at all.

A great example of this type of fraud occurred in a small town in Washington County in the late 1960s. W. S. Brown learned about the case around 1979 while interviewing Gordon Foote, a long-time resident of the town. One of the key people involved in the case was Foote's friend Zeke Foster, a retired state trooper who had hunted timber rattlesnakes with him in New York and Vermont during the 1950s, '60s, and '70s. As the story goes, a man from New Jersey (the same man who served time in a federal prison in 1993 for illegally trafficking in timber rattlesnakes in the northeast) arrived in town in the summer of 1969 with 121 dead timber rattlesnakes, went to the town clerk's office, and submitted a claim for $605.00. In the paperwork that was mandatory for bounty hunters to fill out, he claimed that he had killed the snakes in Washington County. Something about this man aroused the clerk's suspicion, so she gave Foster a call (he had retired a few years earlier but lived nearby), and asked him if he could come over to the office to verify a rattlesnake bounty claim. The first thing Foster did upon arriving shortly afterward was to ask the New Jersey man to take him on a ride and show him exactly where he had killed the snakes. As they drove out of town to a remote and mountainous area a few miles away, little did the man know that Foster knew every den in the county like the back of his hand.

After arriving in what might have looked like promising timber rattlesnake habitat to the man from New Jersey, he suddenly pointed to a ridge off in the distance and blurted out that that was where he had killed the snakes. Foster knew there were no timber rattlers anywhere near that particular ridge, and told the man, according to Brown's account, "You're a god damned liar." They thereupon drove back to the clerk's office and the claim was rejected. Brown later discovered a copy of the rejected claim in the Washington County financial ledgers while searching old bounty payment records, and made a copy of it for his own records. Among other things, it gives the poacher's name, the year, and a claim for $605.00 for 121 snakes. The most interesting aspect of the claim from a historical perspective is that it was crossed out with a single ballpoint pen line, indicating clearly that it had been rejected. It also dovetails perfectly with Zeke Foster's account of the attempted fraud.[27]

The best example of out-of-state residents other than Vermonters attempting to defraud the timber rattlesnake bounty system in northeastern New York was related to me by Pat Leclaire, a town clerk in Washington County in the summer of 2005.[28] Although Leclaire remembers many critical details of this attempted fraud, she cannot pin down the exact year in which it took place. All she remembers is that it took place in the town just to the south of hers not long before she took office in 1972. The case is similar to that in which the claim for $605.00 was denied to the man from New Jersey in 1969. It too involved a perpetrator (or in this instance, perpetrators) from New Jersey, and its outcome also hinged on the decisive intervention of Zeke Foster.

As the story goes, one summer afternoon in either 1970 or 1971, two men with New Jersey plates on their vehicle arrived at a town clerk's office in Washington County with a box full of frozen timber rattlesnakes they claimed to have killed in the county. The clerk may well have become suspicious of the two men when they presumably gave a local address when filling out the necessary bounty paperwork. The law in Washington County as of the passing of Resolution No. 41 on February 19, 1970, was that claims would only be paid on timber rattlesnakes killed within the county by residents of the county. The clerk was apparently quite busy that day, and instead of taking the time to confirm the number of snakes in the claim, asked the men to leave the box of snakes on her front porch on their way out. The New Jersey men might well have thought at that point that their bounty claim would be approved, and that they would eventually receive a check from the Washington County Treasurer's Office, but this is not the way events transpired.

When the clerk, whose name was Marion Guerin, stepped out onto her porch several hours afterwards, she was stopped dead in her tracks by what she described as a buzzing sound from inside the box. The town clerks in New York were used to disposing of the tails and rattles of dead rattlesnakes, since these were the only parts of a timber rattlesnake that a bounty hunter had to produce when making a claim. Marion Guerin had never had to deal with a box of live snakes before, so she crossed the street to the local police station and informed the officer at the desk that some bounty hunters had left off a big box of frozen rattlesnakes, and that they had somehow come to life and were buzzing.[29] The policeman didn't know how to deal with the situation any more than Guerin did. After discussing the situation briefly, they both realized there was only one thing to do, and that was to call in Zeke Foster.

When Foster came, he decided that the best way to dispose of the snakes was to dig a deep hole, burn the snakes in the hole, then cover them up with dirt. I asked Pat Leclaire to tell me as much as she could remember about the digging and whether Foster killed the snakes before he burned them, or if the burning itself was their cause of death. It is now approximately thirty-five years since that afternoon, and she cannot remember exactly how the snakes were killed, other than the fact that kerosene or gasoline was poured on the box after it was thrown down into the hole, and ignited. She is much more precise in her memory, however, about the depth of the hole that was dug. Pat claims it was somewhere in the neighborhood of five to six feet deep. I was quite skeptical but became much less so when she explained that the Fosters lived just down the road from her, and that her two sons helped Zeke dig the hole. I asked Leclaire why Foster had dug such a deep hole, and she remembers him saying that he was afraid that some animal might attempt to dig the snakes up if the hole were too shallow. Zeke had dug the hole on the edge of a large field he owned where many of the children in the neighborhood played baseball. His concern was that a child might possibly step on a snake's head and be pricked by a fang. Even though such an injury would not be a serious medical situation, Foster didn't want to take the risk of injuring or traumatizing a child in any manner.

Once the rattlesnakes were disposed of, Foster apparently decided to find out more about the two men from New Jersey and their very unusual claim. According to Pat Leclaire, he was able to locate the men, who were staying in the area, and talk with them about where they had killed the snakes, why they had frozen them, etc. Unsatisfied with their answers, he took it on himself to conduct an investigation. Leclaire surmises that it was probably through Foster's position as a former state trooper and his police connections in New Jersey that he was able to find out some very incriminating evidence about the two individuals. Apparently, in Leclaire's words, they were in the business of capturing pregnant female timber rattlesnakes in their area (presumably northern New Jersey and southern New York State), letting them have their litters in captivity, then freezing the newborn snakes along with their mothers. Once they had accumulated enough snakes to make the trip worthwhile, they would bring them to clerks in Warren and Washington counties and turn the mothers and babies in for five dollars each.

Today there is no way of knowing just how many times they attempted to get away with this fraud, but they didn't get away with it in this particular instance. Bounty claims were only paid when the county boards of supervisors approved the paperwork sent to them by the various town clerks in their counties. Once Zeke Foster knew how shady these two perpetrators were, it wasn't hard for him to stifle their claim. Leclaire told me that he was held in considerable respect in the town. If he explained this claim and his subsequent investigation to the town's supervisor, which Leclaire feels he must have done, it's very likely that it was rejected before even making it to the Washington County Board of Supervisors for approval.

There is another way in which some individuals may have seriously defrauded the timber rattlesnake bounty hunting system in northeastern New York. If it did in fact take place, it was similar to some of the other frauds I have discussed but also uniquely different. W. S. Brown, Randy Stechert, Steve Harwig (one of Pennsylvania's top timber rattlesnake experts), and others harbor a strong suspicion that at least one New Yorker routinely purchased timber rattlers at Pennsylvania rattlesnake round-ups (the Morris round-up in Tioga County and others) for as low as a dollar each, then took them up to Warren and Washington counties in New York to collect a five-dollar bounty on them during the 1960s and 1970s.

Harwig wrote to Brown on August 30, 1981, detailing a number of facts and incidents that, when viewed as a whole, had made him suspect this particular fraud. The first of these took place in the mid to late 1960s, a period of time in which Harwig attended the Morris round-up in Tioga County on a number of occasions. On the morning of June 5, 1965, while stopped at a general store just outside Morris, he was told that a man with New York plates and a lot of rattlesnakes in the back of his truck had just gassed up and headed north out of town. Then, a few years later he discovered that the officials of the Morris round-up had somehow lost their records for the years 1964 through 1966. It occurred to him that something fishy may have taken place, but there was no way of knowing exactly what it might have been.

As Harwig explained in the letter, he apparently didn't suspect any New Yorkers of possible interstate shenanigans with timber rattlesnakes until after studying some of the available timber rattlesnake bounty payment records from Warren and Washington counties in New York that W. S. Brown had sent to him years later. The first thing that seemed very odd to him was the fact that Warren County had a gap in its timber rattlesnake claims in the exact same years that the Morris round-up had a gap in its records. Was this sheer coincidence? He made it clear that he didn't think so. His suspicion was that somebody in Warren County must have known that something funny was going on with the rattlesnake bounty claims in the years 1964 through 1966, and that this was why the records for those years were conveniently lost. Another thing that bothered Harwig was the fact that Warren County paid only an average of forty-seven timber rattlesnake bounty claims per year from 1954 through 1963, compared to an amazing average of 390 claims per year on the snakes from 1967 through 1973. What could possibly account for such a tremendous jump in claims? He went on to express his suspicion that a good many of the snakes that were claimed in Warren County in the 1960s and '70s were "imports" that had either been killed or purchased in Pennsylvania. He also pointed out in his letter that this fraud probably started in the Morris round-up gap years of 1964 through 1966, and that this may have had something to do with the county's mysterious loss of its rattlesnake bounty claims records for those years.

One of the things that Steve Harwig did not emphasize in his 1981 letter to W. S. Brown, even though Brown had sent him fairly extensive bounty payment records from both Warren and Washington counties, is the fact that Washington County also had an inexplicable gap in its bounty payment records in the mid-1960s (1964–1967). In addition to this remarkable coincidence,

Washington County paid far more timber rattlesnake claims per year from 1968 to 1973 than it did from 1953 to 1963. From 1953 through 1963 it paid an average of 273 claims a year, as compared to an average of 439 claims a year from 1968 through 1973. That's a sizable increase. "Why" is a mystery that will never be solved.

We'll never know for certain if an individual or team of individuals perpetrated and got away with this particular fraud, but there certainly is some strong circumstantial evidence pointing to this possibility. An important thing to bear in mind about bounty systems is that they are notorious for fraud. It would have been truly remarkable if various bounty hunters and other individuals had not figured out ways to get around the various restrictions within the timber rattlesnake bounty systems in northeastern New York during the middle years of the 1900s when the bounty on the snakes was as high as five dollars per snake in Warren and Washington counties. Despite the fact that Essex County had a five-dollar bounty on timber rattlers from 1956 through 1971, I don't believe that anyone attempted to defraud the system there, since its relatively small overall population of timber rattlers would have made any overly productive rattlesnake bounty hunters fairly easy to detect.

With Rutland County rattlesnakes being turned in for the higher bounties in Warren and Washington counties, we'll never know just how many timber rattlers were actually killed in Rutland County, especially during the 1950s, '60s, and '70s. Undoubtedly, many more Vermont rattlesnakes were killed during those years than any of the available records would indicate. If anything, this reinforces my belief that bounty hunting in Rutland County seriously depleted the state's timber rattlesnake population. On the other hand, the deflation of Vermont's numbers by border-jumping means an inflation of New York's numbers.

Another thing that undoubtedly inflated New York's numbers from the early 1950s through the early 1970s was the possibly sizable number of timber rattlesnakes from New Jersey, Pennsylvania, and probably other parts of New York that were falsely claimed in Warren and Washington counties. Regardless of the fact that the number of claims in New York was inflated by the shenanigans of a number of good old boys trying to make as much money as possible on the rattlesnakes they killed, in no way should it be assumed that Art Moore and other bounty hunters didn't significantly deplete the timber rattlesnakes of Warren and Washington counties. It probably wouldn't have been as significant a depletion if timber rattlesnakes on the northern fringe of their range were capable of rebounding quickly from depletions. We know, however, through W. S. Brown's long-term field study that this is anything but the case. This species is capable of maintaining its numbers on the northern fringe of its range as long as it is undisturbed, but it can take decades to increase its numbers even marginally.

CHAPTER 7 : A FAMILY TRADITION
Lester and Carol Reed

As previously mentioned, Lester Reed and his daughter, Carol, were two of Vermont's best known timber rattlesnake bounty hunters. Their heyday of hunting the snakes together was the mid-1960s through the mid-1980s. Although Vermont discontinued its bounty on timber rattlers in 1971, Lester and Carol continued to hunt the snakes through 1986, the last year the snakes could legally be killed or captured in Vermont before they were added to the state's endangered species list. They stood out in their small town in western Rutland County not only because of the fact that they hunted together as a father-daughter team, but also because of the locally famous timber rattlesnake oil that their family had produced for medicinal purposes since the early part of the twentieth century.

Timber rattlesnakes and the Reed family go back a long way together. It all started back in the late 1800s on Lester's grandparents' farm on the outskirts of town. Nestled on the southern end of a long ridge, which was and still is well known for its timber rattlesnakes, the old homestead was built right on a migration route commonly used by the snakes. Lester's grandparents, Frank and Clarinda Isiola Waterhouse, saw and killed rattlers on their property every summer, and they were the first ancestors in Lester's memory to have processed rattlesnake oil.

When Lester's father, Harry, married the Waterhouse's daughter, Lillian, the stage was set for a serious rattlesnake hunter to emerge. Harry was a natural to fill the role. Born in 1902, he grew up in a life style that was a throwback to the previous century. His family basically lived off the land, knew how to work hard, and did whatever it took to make ends meet. Despite growing up poor on a small farm, getting up early in the morning, and having to work long hours even as a teenager, Harry was just like his father (Lester's grandfather) in one respect. Whenever he had the time, he went hunting or fishing. It was more than something fun to do. It put food on the table. During his teens Harry developed into an accomplished outdoorsman under the tutelage of his father. There wasn't anything he couldn't do successfully in the out of doors, even trapping foxes.

It was no fluke that Harry took up rattlesnake hunting after his marriage in the 1920s. He had the main ingredients he would need to be successful. He was a good outdoorsman and hunter, he was poor, and he needed all the extra money he could come by. The one dollar bounty on timber rattlers in Vermont must have attracted him. A dollar meant a lot more in those days than it does now, and he wasn't afraid to work hard for a dollar. He also had the desire to start hunting the snakes after marrying Lillian and becoming a member of a rattlesnake oil-producing family. Since none of the Waterhouses were bounty hunters, and only killed the snakes when they had the opportunity to do so on their farm, young Harry was more than willing to start providing them with as many rattlesnakes as he could. The fact that he lived only two and a half miles from the nearest rattlesnake den must have been a factor too. A walk of that distance was nothing to Harry. He had walked almost ten miles a day round-trip to hold down his full-time job in a slate quarry when he was only twelve years old.

During the 1920s and 1930s Harry became highly skilled as a rattlesnake hunter. With his broad skills as an outdoorsman, it was bound to happen. He was constantly bringing home a wide variety of game. Whenever he took to the woods or marshes, he'd return home with something to eat or pelts to sell. He and Lillian ate rabbits, squirrels, woodchucks, turtles, frogs, ruffed grouse, and deer. They also enjoyed bullheads, perch, and northern pike from a big marsh near their home. It was no surprise that Harry consistently brought home timber rattlesnakes for oil production during those years.

In 1936 Lester was ten years old and following in his father's footsteps in the out of doors. He remembers going up to the rattlesnake den by the late 1930s as an observer. Ironically, as accomplished as Lester later became, Harry had real doubts about his son ever becoming a timber rattlesnake hunter. Lester apparently wasn't able to pick the snakes up visually as well as his father could. He admits that it took him a long time to acquire that skill, and recognizes that it's one of the essential skills needed to hunt them. As mentioned earlier, you can be within two or three feet of timber rattlesnakes and never notice them unless they move or rattle. Their camouflage among rocks or on a hardwood forest floor is exceptional.

With Harry as his mentor, Lester was destined to become an all-round outdoorsman and rattlesnake hunter. The Reeds were effective at passing on their way of life and their skills from generation to generation. It was the way they survived. It's a little hard to imagine a person working as hard as Harry Reed, but Lester proved to be a chip off the old block. In the summers as a boy he worked in an apple orchard, sometimes putting in ten-hour days. He also mowed grass for one of the town's more prominent families. Unlike his father, who quit school at the age of twelve to go to work full time in a slate quarry, Lester stayed in school through the eleventh grade, so most of his early working career was limited to the summer months. That pattern changed when he got his driver's license at age sixteen and went to work for a small local dairy. During the school year he'd deliver milk to the stores in town early in the morning, go to school all day, then go back to the dairy in the late afternoon to clean up and do chores. In the summers after getting his license, he worked at the dairy full time.

As busy as he stayed with school and his jobs, he developed into a topnotch outdoorsman on the weekends during his teenage years. By the time he married his wife, Theresa, at age twenty-three, he was working for the town's major dairy and was a seasoned hunter, fisherman, and trapper. He had also developed the ability to see rattlesnakes in well-camouflaged positions during his teens, and become very effective as a timber rattlesnake hunter, often teaming up with his father for hunts at the den north of town. By the early 1950s Lester and Harry had come into their prime as a timber rattlesnake hunting team, and were bringing in more than enough of the snakes to keep Lillian Reed busy with her rattlesnake oil production. Lillian, Lester's mother, was the family member in charge of processing the oil and took her position seriously.

Lester and his father had a few great years hunting rattlesnakes together before Harry suddenly died in the summer of 1955. For nearly ten years afterward Lester continued to hunt the den, but it wasn't the same without his mentor and companion. He often felt lonesome up on the talus slope and there were days when he thought he'd be solo hunting the snakes for the rest of his career. How wrong he was. He had overlooked his daughter, Carol, who, like numerous Reeds before her, loved hunting and fishing early in life. In 1964 when she was in the eighth grade, she accompanied Lester on her first timber rattlesnake hunt and was hooked immediately.

To understand why Lester decided to take Carol up to the den in the first place, several factors have to be considered. First of all, Carol wasn't your run-of-the-mill tomboy. She was an exceptional tomboy. Lester always wanted a

FIGURE 5. Rattlesnake bounty hunter Lester Reed reflecting on a timber rattlesnake den. Photo taken in September 2004 by Jon Furman.

boy, according to his wife, but in Carol he got a girl and a boy. She was a brave child who was not afraid to try things that might be considered a little daring for the girls of her era, including driving go-carts, water skiing, playing baseball, and getting into touch football games with the neighborhood boys. Lester really admired her grit.

In addition to her courage, Carol was a remarkable little outdoorswoman. The fact that there were very few females in her area that hunted didn't stop her from following in the Reed tradition of becoming a good hunter. Nor did it stop Lester from being a great mentor for her. Carol hunted everything with her father from bullfrogs to whitetails. She was only eight years old when she shot her first buck. Lester had shortened the stock on an old .308 caliber rifle and passed it on to her. It kicked pretty hard, but Carol became deadly accurate with it. She had been shooting a .22 rifle with a sawed-off stock since the tender age of five and had become a real sharpshooter. When she saw her first buck coming down a deer run north of town, she raised her gun, fired, and dropped him at about sixty yards.

It seems likely that Lester would never have taken Carol rattlesnake hunting based on her grit and hunting ability alone. There were two other factors that made him decide to take her on her first snake hunt. The first of these was the remarkable bonding that they shared. A prime example of this was the fishing that they so much enjoyed. All year round they fished together for bullheads, perch, and northern pike on the big marsh past the Waterhouse family homestead. Even on the coldest winter days they'd ice-fish on the marsh and think nothing of it. It didn't really matter what they were doing in the out-of-doors. As long as they were together, they were happy.

The major factor that broke Lester down to the point where he was willing to take Carol up into timber rattlesnake country was the pressure she exerted on him to do so. Parents are known to dote on only-children, and only-children often have their parents wrapped around their fingers. This appears to have been the case with Lester and Carol Reed. When I asked Theresa Reed if Carol had Lester wrapped around her finger as a kid, she told me candidly, "You don't know the half of it." When I asked her if Carol had pressured Lester into taking her rattlesnake hunting, she told me that she had, and that that's what made the difference. Otherwise he would probably never have taken her, in spite of the fact that he wanted to share the excitement of the sport with her.

The first time Lester took Carol rattlesnake hunting he couldn't stop worrying about her safety. Even though he had warned her to be careful and alert from the time they entered the woods below the den until they left the woods later, he knew there was a chance she could be hurt. What he feared more than anything else was a bad fall. As it turned out, things couldn't have gone better. Carol moved around well on the steep, boulder-strewn slope, and stayed focused throughout the hunt. Lester located and shot three large timber rattlers on that late Indian summer afternoon. All in all it was a perfect introduction for Carol into the realm of the timber rattlesnake and the sport her father loved above all others.

Everything was perfect until the two hunters arrived home just before supper with the three rattlesnakes and a sheepish look on their faces. The one thing they had neglected to do before heading off at three-thirty was to tell Lillian, who had lived with them since her husband's death in 1955, the truth about what they intended to do. They knew she would try to stop them from going rattlesnake hunting, so they told her they were going squirrel hunting instead. When they returned, she was extremely upset at Lester not only for deceiving her, but also for placing Carol in harm's way. She'd had a long track record of being worried every time Lester and his father went rattlesnake hunting, and the thought of her granddaughter taking up the sport was something she didn't want to entertain at all. Being as outspoken as she was, she had no trouble voicing her displeasure. Theresa was almost as upset, but she elected to give Lester the silent treatment instead.

Despite a little negative feedback, the horse was out of the barn and there was no stopping it. Carol couldn't wait to go rattlesnake hunting again, and Lester ended up taking her three more times that fall. During those hunts he realized he had found an ideal partner. Carol not only loved being up on the steep slope surrounding the den, she had sharper eyesight and better hearing than he did. Although she was in the apprentice mode that first year, he knew it wouldn't be long before she became a full-fledged rattlesnake hunter. She had done well in all the other areas of hunting he had introduced her to, and everything indicated that she'd do well in this one too.

Within another year or two the new Reed rattlesnake hunting team came into its own. Where the other timber rattlesnake bounty hunters they knew tended to hunt the snakes in the spring, Lester and Carol did the brunt of their hunting in the fall. Their goal was not to kill as many rattlesnakes as possible, and in that sense they were not typical bounty hunters. They also didn't bother to collect the bounty on all the snakes they killed. If they only had two or three snakes after a hunt, they often went home without stopping at the local town clerk's office to make a bounty claim. Their primary goal was to harvest enough snakes each year to produce their oil. Experience had shown Lester, and his father, Harry, before him, that the best time of year to hunt the snakes for oil production is in the fall.

There are two fat deposits in the body cavities of timber rattlesnakes that are at their maximum size in the fall, just before the snakes go into hibernation. As mentioned in a previous chapter, it's these deposits of fat from which rattlesnake oil is produced. Lester claims that they needed somewhere in the neighborhood of a dozen snakes to produce six ounces of their oil each year, or two large timber rattlesnakes to make an ounce of oil. Smaller snakes, of course, have less fat, and there are times when large snakes have much less fat than might be expected. So the number of snakes they needed to produce their oil annually depended on the fat content of each timber rattler they killed.

Knowing how much this fat meant to them, I asked Lester what kinds of weapons they used to dispatch their snakes as quickly as possible without damaging the all-important fat deposits. In Lester's case, he used a 16-gauge shotgun and number 4 shot. It's the gun he took along the first day he actually hunted rattlers with his father back in the 1940s, and he stayed with it throughout his career. The Reeds' hometown has its share of classic Vermonters, and Lester is definitely one of them. "If it ain't broke, don't fix it" is one of his mottos. The reality is, though, that a blast from a 16-gauge shotgun could blow a rattlesnake to smithereens. So how did he avoid damaging the all-important fat deposits? "It was a little hard sometimes on coiled snakes, but I was pretty good at homing in on their heads. I seldom hit their fat at all," he explained to me. Even if he was deadly accurate for the most part, his choice of a weapon was questionable. A shotgun on the slope he hunted gave him plenty of killing power, but it also tied up one or both of his hands at all times, since he never put a sling on it. With a pistol loaded with birdshot and carried in a holster he could have had both hands free at all times except when actually shooting a snake or dealing with a dead snake. Had he hunted this way, he would have been better able to prevent or break falls.

Lester had one very bad fall in his career and was lucky not to have been killed or seriously maimed. It took place one evening when he stepped onto a yellow jacket ground nest while coming down alone from the den. Immediately he knew he was in trouble and started to run downhill. Within seconds he was out of control. Fortunately, he was able to grab onto a small sapling in the last millisecond before disaster struck. He grabbed it with his one free hand

and spun so violently around the tree that it tore the skin off the inside of his arm from his biceps all the way down to his wrist. Over fifty years later he still has the scars from that injury. If he had gone into that fall with both hands free instead of only one, while his other hand held his shotgun, he might have been able to grab the sapling with both hands and minimize his injury considerably. Amazingly, he never dropped his shotgun despite the severity of his fall and being stung about two dozen times.

Carol, luckily, never had a fall. She started to lose her balance on numerous occasions, and several times she started to slide downhill, but her saving grace was that she wore her weapon in a holster on her hip. It was a nine-shot .22 revolver loaded with birdshot. She was always able to stop a possibly serious situation before it got out of control by grabbing onto a sapling with both hands. Saplings weren't common up on the slope, but there was always one nearby when she needed one.

Accidents were something Lester and Carol did their best to avoid. They developed a pattern of starting all their hunts approximately in the middle of the den. The den is fairly wide and stretches over seventy-five yards or so of talus. From there they moved across to the northernmost part of the den and eventually arced back across the slope, ending up at the southernmost part of the den. The reasons for this pattern were twofold. They always hunted the middle of the den first because it was closest to the road, and they always hunted the southernmost part of the den last because it offered them the safest descent from the ridge. The slope below it was less steep and had a number of saplings during the era in which they did the bulk of their hunting together, the late 1960s and 1970s.

I asked them if they ever had any close calls with a timber rattlesnake, and each of them, as it turned out, had one terrifying experience that will stay in their memories until the day they die. One day Lester was coming down from the southernmost part of the den after a solo hunt, and it was a scorcher of a day. It was so hot in fact that he did something he had rarely done in the past, and never did afterward. He stopped for a drink of water at a place near the bottom of the ridge where water flows out of a spring and forms a tiny rivulet about a foot or so wide. As he crouched, cupped his hands in the water, and raised his hands to his face, he saw a flash out of the corner of his eye. It was a small rattlesnake, approximately twenty inches long, striking at him. Fortunately it missed him by several inches. It had been on a rock on the other side of the water, and its momentum had carried it completely onto his side. With a surge of adrenalin, Lester grabbed a stout walking stick, which he had carried that day in addition to his shotgun, and clubbed the snake to death. Afterward he was shaking so badly that he couldn't even drink. As Lester looks back on the incident, he assumes that the snake must have been striking defensively at his movement.

Carol's close call came one fall afternoon while hunting with Lester. They were about three-quarters of the way through their normal hunt, and basically on their way down to the road. She had just stepped from one rock to another and paused when she heard a little tapping or thumping sound. It had what Carol described as a hollowness to it. She looked around to see where it was coming from. Then she looked down and realized she was standing right on the middle of a rattlesnake. It was a small snake, about twenty-four inches long they later determined, and it was striking at her boot. Instinct took over. In an instant she lifted her leg, drew her gun, and shot the snake in the head before her foot even touched the ground. Lester, who was standing nearby at the time was surprised by how fast everything had happened, and so was Carol. Had that timber rattler been a forty-eight-inch-long, two-and-a-half-pound male, the situation might well have turned out differently. Carol was lucky that afternoon.

With the Reeds' reputation in their town and elsewhere in the western part of Rutland County, it seemed likely that over the years they must have taken out some amateur herpetologists, thrill seekers, or people curious about timber rattlesnakes. They told me that they'd taken out about a dozen individuals in their careers, but only a few of them stand out in their memory, such as a man who came to them in the 1960s. He was a Vermonter who totally doubted the existence of timber rattlers in the state. Apparently, he was a quiet man for the most part and a real Doubting Thomas type. Lester agreed to take him out, hoping they would encounter at least one good-sized snake during their field trip.

Lester had seen a variety of reactions from people he had taken up into rattlesnake country, all of which mildly amused him. These included turning as stiff as a board and sweating profusely in the presence of a buzzing rattlesnake. One reaction he had never seen before occurred the day he took out the Doubting Thomas. It had been a long hunt across the talus slope without any action whatsoever. Lester was beginning to lose hope. It was summer and not a good time of year to find snakes anywhere near the den area. The two men paused momentarily on a large rock to catch their breath. Then the doubter stepped forward and down, and what his body accomplished in the next half second totally defied the laws of physics.

Lester swears the man jumped ten feet through the air. As soon as his feet touched the ground, he was airborne. There was only one thing on that slope capable of causing that kind of reaction, and Lester knew exactly what it was. Sure enough, a large, coiled timber rattler was within inches of where the man's feet had landed initially. When Lester's visibly shaken guest was composed enough to talk, he said something like, "OK, I've seen enough. Let's get out of here." It's not every day that one gets a chance in life to present such overwhelming evidence to a chronic doubter. For that reason alone the trip was well worth the effort from Lester's point of view. How big an impact did the rattlesnake have on his guest? Lester may have summarized it perfectly. "He was a pretty quiet individual when he arrived that day and a lot quieter when he left."

In August 1965 a man named Kinsman Lyon from the Museum of Science in Boston showed up unexpectedly at the Reeds' back door. He said he had heard of timber rattlesnakes existing in western Rutland County, had come up to fact-find, and been given Lester's name. Lester told Mr. Lyon he had come at a bad time of the year. There'd be no snakes on the den the family hunted. To accommodate the man, he drove him north to show him the general location of the den and invited him to return in October for a fall hunt.

Time went by and the Reeds wondered if they'd ever hear from Kinsman Lyon again. He had seemed pretty skeptical about things the day he left in August. Finally, in late September, he called one evening and arranged to go on a hunt with them as an observer. About a week later he arrived for the hunt and asked Lester and Carol if they could try to capture a live snake for him. He obviously was looking for some indisputable proof of the species existing that far north in New England. It turned out to be an eventful afternoon. Lester and Carol captured a young rattlesnake, shot two or three large rattlers, and their guest was nearly scared out of his pants on two different occasions. In the first of these, one of the Reeds startled an adult rattler which slithered right between his boots on its way to a crevice. In the second situation Carol shot a large adult within three feet of him after which he said, "I'm awfully glad you know how to shoot." When they got back to the house, Mr. Lyon had clearly had enough. Carol remembers him saying, "You've certainly proven what I was looking for." He thoroughly thanked his guides and drove off, never to be heard from again.

The fact that there are and always have been people who doubt the existence of timber rattlesnakes in Vermont is somewhat amusing to anyone who knows better. All any doubter would have to do is call or go to the various town clerks in the western part of Rutland County. If they are persistent enough, eventually they are going to come across some old records of timber rattler bounty payments. They could also call Vermont's Fish and Wildlife Department and ask them about the snakes' current status in the state. Even though the Fish and Wildlife Department would justifiably not be likely to pinpoint the exact location of timber rattlesnake dens in Vermont, they would not deny that remnant populations of the snakes do exist in Rutland County. I sometimes get the feeling that chronic doubters, who are often adamant about their beliefs, don't want to learn the truth.

It's probably safe to assume that many of the people who doubt the existence of timber rattlesnakes in Vermont or in the eastern foothills of the Adirondacks might also doubt that rattlesnake oil has any kind of medicinal benefit whatsoever. They'd tend to be close-minded about the subject. The Reed family never doubted the medicinal uses of rattlesnake oil. They used it for burns, earaches, cuts, and painful arthritis, among other ailments. There was a period of fifty years or more when they never dreamed of being without rattlesnake oil, or in short supply of it. In fact, they still have some small bottles of the golden oil from the late 1970s, since it stores so well.

The way Lillian produced the oil was quite a process. First she'd remove the fat deposits, which are wrapped in a delicate membrane, from a timber rattler's body cavity. Depending on how clean they were, she'd soak them in lukewarm water with a little salt for up to half an hour. If they were relatively clean and free of blood, she might only soak them for five or so minutes. Next she would work the fat out of the membrane by thumb into a preheated aluminum pan. At this point she'd keep her eye on the process to make sure the fat didn't crystallize. Eventually a golden oil would begin to seep out of the fat. To get out the last remnants of oil, she'd often work the fat with a butter knife. Once she had extracted as much oil as possible, there was one final step. She'd then funnel it through cotton batting into a glass jar to remove any tiny impurities. The whole process generally took about an hour. How important was the end product to Lillian? Carol told me, "The oil was like gold to Nan."

One day when Lester was a boy he learned the hard way how seriously his mother took her rattlesnake oil. She had processed some oil to the point where the only thing it still needed was to be filtered into a glass jar. Apparently she didn't have time to complete the final step, so she poured it into an earthenware bowl and put it on a shelf that was built into the back of her cast iron cook stove. She frequently put breads and muffins there to keep them warm. Sensing no harm, she left the house for a while, perhaps to visit her parents up the road. When she returned later, the oil had seeped through the bowl and was all over the cooking surface of the stove. Upon seeing the mess, she must have assumed that Lester had spilled the oil. She couldn't initially fathom that her rattlesnake oil, which was known for its penetrating ability, could actually have penetrated through the bowl, and yet it had. Lester remembers being outside when the door flew open and she lit into him with a serious tongue-lashing. It took him quite a while to calm her down and convince her of his innocence. Carol told me that wasting the oil that day was as bad a blow to her grandmother as losing a fifty dollar bill. Considering this episode took place in the heart of the depression, that's saying a lot.

Lester, Theresa, and Carol Reed have some extremely convincing examples and stories of how their family successfully used rattlesnake oil over the years. It's natural to have a high opinion of one's own product, but it's more convincing when somebody else has a high opinion of it. Were there other people in town who believed in the Reeds' snake oil? Yes, there were, but nobody more so than a man named Erwin Mallory. Back in the 1960s he was constantly hobbled by painful arthritis in one of his knees. It was just before the era of widespread artificial knee and hip replacements. As his condition intensified, he apparently reached the point where he was willing to try any remedy that might help him. So, one day he arrived at the Reeds' house and asked them for a little jar of their snake oil. Several weeks went by without a word; then one evening he returned, walking almost normally and wanting more. One little jar had made him a believer.

The Reed family had a remarkable involvement with Vermont's timber rattlesnakes for over half a century. It started back in the 1920s when Lester's father, Harry, began to hunt them both for the bounty and for their fat. There's no doubt in my mind that if the state had never included the snakes on its endangered species list in the late 1980s, there would be one or more members of the family still hunting their favorite ridge on warm afternoons in late September and early October every year. It's now been about twenty years since Lester and Carol last hunted at the den they came to know so well. I asked them if they miss their rattlesnake hunting, and they both immediately indicated that they do. Perhaps more than anything else, they miss the hair-raising thrill of the sport. There's something inherently exciting about being up on talus slopes in search of these snakes. It's definitely an adrenalin rush to hear a rattlesnake buzzing in agitation or warning, especially when it's nearby but out of sight. On a normal hunt Lester and Carol were on constant alert from the time they entered the woods below the den until they left them later, often with two or three large rattlesnakes in a special bag Lillian made for them. There was never a dull moment. In their case they were bound to become diehard, high-level rattlesnake hunters because of the excitement and bonding it gave them, their love of hunting, and their family tradition of producing rattlesnake oil. Only a handful of rattlesnake hunters in Vermont history were as committed to the sport as they were.[1]

CHAPTER 8 : A MODERN DAY PIONEER
Bill Galick

I'll always remember the late winter afternoon when I first approached the old Galick farm in western Rutland County, on an endlessly long driveway, with a little feeling of trepidation. I was on my way to meet Bill Galick, who had agreed to be interviewed about his timber rattlesnake bounty hunting days. Although he had seemed amiable enough on the phone, some things I had heard about the Galicks over a thirty-year period of time—men on their farm wearing six-guns on their hips—plus the incredible ruggedness and remoteness of their property, had me on edge. Despite my anxiety, there was no turning back. Anyone writing about timber rattlesnake bounty hunting in Vermont would have to seek out and include Bill Galick, who was well known not only for his bounty hunting and countless tales of rattlesnakes on his property but also for having one of the only remaining active timber rattlesnake dens in the state.[1]

It didn't take long to realize that first afternoon that Bill was a tough, independent old hombre. He asked me to sit down across from him at his kitchen table and proceeded to maintain a total poker face for the next several hours as we discussed timber rattlesnakes and his long career as a timber rattlesnake bounty hunter. During the first hour of that visit he never had his hand more than six inches away from a loaded revolver, which at first was hard to detect on the liquor-and-pill-bottle-strewn table between us. He was in his eighties, about as far back in the country as it's possible to get in Vermont, and determined to show me that he was capable of defending himself.

It took a number of visits to really break the ice with Bill. Once he realized that I meant him no harm and was genuinely interested in his past, he gradually told me more and more about the remarkable lifestyle he had lived on his nearly two-thousand-acre farm for as long as he could remember. One of the most interesting facts that emerged from our many discussions over a

FIGURE 6. Rattlesnake bounty hunter Bill Galick, circa 1996.
Photo taken by Jennifer Hartman.

long period of time was the extent to which his background overlapped with Lester Reed's. Both men grew up poor, and had to work long hard hours as boys. To a large extent their release in life was in the out-of-doors where they trapped, hunted, and fished. They didn't take to the woods, nearby ponds, and marshes for sport alone. The game they brought home and the furs they sold represented a significant part of their family's overall survival.

Bill Galick's family came to their farm from Schenectedy, New York, in the winter of 1919. The original settlers were Bill's grandparents, his parents, five uncles, an older sister, and himself, an infant at the time. For their first ten or so years in Vermont the family basically lived off the land. They raised animals for slaughter, grew crops, and gardened. The closest town of any size was in New York State, so they did the bulk of their business and got most of their provisions there instead of in Vermont. In their early years in Vermont, the Galicks had no driveway. Since it took them longer to get over to New York via Vermont roads than by crossing Lake Champlain, their norm was to drive across the lake in the winter when the ice was thick enough, and take boats the rest of the year. This is how Bill, his two brothers, and four sisters went to school as well. Even though they were residents of Vermont, they attended school in New York.

By the early 1930s and Bill's teens, his father and uncles were dairy farming, milking over fifty cows. He may have had it relatively easy as a young boy, but now his hard work began in earnest. I asked him what was the hardest job he could remember during that period of time. He didn't have to think long before responding, "It was the haying." He claimed that his family had the first hay truck in Vermont. Bill's father, John, and his uncle, Steve, were extremely adept mechanically, with excellent engineering and welding skills. They had a complete machine shop in one of the farm's many barns, as well as welding equipment, and could make and/or alter just about anything. The hay truck they built was a prime example of their resourcefulness. It was a modified Buick automobile with an eighteen-foot-long bed mounted to its frame. It was on the bed of that truck that Bill worked like a man every summer from age thirteen until he went into the Army and World War II.

As the Galicks hayed, they dragged a hay loader behind their truck which threw loose hay up onto the truck's bed. Two men were always on the bed with pitchforks. The rear man threw hay forward to the front man, who tried to stack it up evenly in the front of the bed. Bill often was the front man, but he hated either position. And there was no room for whining or complaining, as his uncles were not opposed to keeping him in line. Besides haying, Bill had to do a wide variety of farm chores including feeding livestock and mucking out the cow barn every day of the year.

It wasn't an easy life for a youngster, but Bill had another world into which he escaped as frequently as possible. The Galick farm had miles of marshes and hardwoods loaded with game. By age fifteen Bill was trapping and hunting voraciously. He was also fishing on Lake Champlain. I asked him how good he was in the woods as a boy and he said, "Damned good. I was in 'em all the time." In the spring he trapped muskrats in the marshes by the lake. In the fall he hunted deer by himself and raccoons with his father in the endless woods on the mountain behind their farm. It was in those days that he must have gotten the idea in his mind that he could live off his land without ever needing to work in the outside world. Even if he wasn't thinking that way, he was becoming proficient as a trapper and learning the ropes of raccoon hunting. Trapping and coon hunting were ultimately two of the skills that allowed him to live off his land and support his family in his adulthood. Another skill was his timber rattlesnake bounty hunting.

I inquired about his bounty hunting and how it all began. The Galicks always knew there were rattlesnakes on their property, and they knew the general location of the den. More than once in his boyhood Bill saw his grandmother kill a big rattlesnake right on their lawn with a hoe. "A dog would start barking and you'd know there was a rattler around," he told me. The snakes had a tendency to get under their front porch on hot summer days. Since timber rattlers feed mostly at night in warm weather, it wasn't uncommon to see them leaving at dusk. In a way, the large wraparound porch on their house was a little like a shelter rock. It certainly was an area the snakes used repeatedly.

By his late teens Bill discovered the den. He knew it had to be up on the property's distant talus slope somewhere. It was just a matter of being up there at the right time and seeing a number of snakes in one area. During his visits to the den prior to World War II the snakes were scarce, no more than a handful at a time, and he doesn't remember ever killing one there. The killing began in 1946, the year after he was discharged from the Army. At that point Bill and his brother Ed realized there was money to be made in bounty hunting. The local town clerk's bounty records for 1946 show that Bill and Ed Galick brought in one hundred and seventeen timber rattlesnakes that year. According to Bill there were a lot of snakes after the war.

The first time Bill went up to hunt rattlers at the den his timing couldn't

have been better. It was early May, and perfect basking weather had finally arrived in the Champlain Valley. Bill could hardly believe the number of snakes
he encountered. "There must have been a hundred of 'em," he told me. "I stood
in one place and counted eighteen." He was wearing hip boots, a practice he
never relinquished for the rest of his career. His weapons were his boyhood
.410 shotgun and a .22 revolver loaded with birdshot. Bill killed twenty-four
snakes that day. He shot a few and, after running out of ammunition, clubbed
the rest with a stick. When he arrived back at the farm some of his relatives
were initially doubtful that he had been so successful. After he dumped out a
sizable pile of heads and rattles from his gear, their skepticism vanished in a
hurry. It was obvious that an emerging bounty hunter was in their midst.

It took Bill several years of hunting at the den to figure out the best weapons
and equipment for his bounty hunting. His .410 shotgun didn't last long. In his
initial hunts it gave him confidence, but it was too cumbersome to carry. Early
in his career he realized that a pistol loaded with birdshot was the best way to
kill timber rattlers, but his .22 wasn't packing enough punch for decisive kills.
For a short period of time he tried a friend's .38 Special and liked it. Then in
the mid-1950s he purchased a Ruger .44 magnum pistol and used it exclusively
afterward. He found that his own hand-loaded birdshot ammunition fired in
that gun packed enormous punch.

One of his most important pieces of equipment other than his powerful
pistol was the snake hook he developed. There were numerous times during
his long bounty hunting career when he needed to grab hold of a rattlesnake
that might otherwise elude him. For this purpose he took a lightweight but
sturdy stick about three and a half feet long and figured out a way to attach a
large treble hook securely to one end. Once he sunk his snake hook into the
side of a rattler, it didn't have a chance of eluding him. Another piece of indis-
pensable equipment was his snake can. It was a bait bucket that he soldered a
clip onto and wore on his belt. He had to have a way to bring back both the
heads and rattles of any snakes he killed, which was the prerequisite for collecting the bounty in Vermont. His can had a lid, was rather deep, and worked
out perfectly for this purpose. It accompanied him on every hunt in his long
career and was a prime example of Galick ingenuity and resourcefulness.
Probably the most important thing about Bill's snake can was that he didn't
have to carry it by hand.

After many conversations with Bill about his past, it was easy to envision
him fully decked out for a hunt and ascending the talus slope on his property.

He wore his hip boots, or trout boots, as he called them, no matter how hot a day it was to prevent rattlesnakes from sensing his body heat and striking at him. On his gun belt he wore his trusted .44 magnum pistol on his right side and his snake can on his left. He always carried his long-handled snake hook in his right hand, leaving his left hand free to prevent or possibly break a serious fall. The terrain he hunted was all too easy to take spills on, which is typical of talus slopes wherever they occur. What was hard to envision was how Bill actually hunted the den. To minimize his chance of falling and seriously hurting himself, he usually walked the same routes across the den, and consequently came to know them extremely well. The Reeds basically did the same thing at their den. Bill occasionally deviated off course to snakes he saw at a distance, but his regular routes usually put him in the most likely areas to see snakes. He often stood in one area and shot several rattlesnakes at a time from a distance of ten feet or less. If they were hit decisively, he was apt to move on, then come back to get them later. A technique he frequently used was to try to stimulate some rattling from snakes he couldn't see as he traversed the den. Very often he would accomplish this by simply standing in front of, or on top of, certain large rocks that often had snakes under them. If he heard any rattling, he'd cautiously look underneath for the agitated snake or snakes. Sometimes he'd be able to shoot immediately, and other times he'd have to go into the hook-pull-and-shoot mode. It's a miracle the man was never bitten.

It's a lucky thing for Bill Galick that all he needed to bring down off the mountain were the heads and rattles of the snakes he killed. The average weight of an adult male timber rattlesnake in the border rattler area of Vermont and New York is about nine hundred to twelve hundred grams, or two to two and a half pounds. I base this on the thousands of timber rattlesnakes that W. S. Brown has weighed in his capture and release field study in northeastern New York over the last quarter century. Brown's data reveal that adult females in his study area typically weigh around one and a half pounds but can weigh as much as two and a half pounds when pregnant. Needless to say, it would have been quite a burden for Bill to come down from his den with a backpack full of adult timber rattlers. He often killed thirty to forty snakes in his hunts, and a load of snakes that large on his back would have been an unpleasant and possibly dangerous weight to carry.

Bill Galick never kept a record of the number of rattlesnakes he killed, but he estimated that he cut the heads and rattles off as many as two thousand snakes in his career without ever accidentally pricking himself with a fang. To accomplish this feat he always carried a small pair of brush cutters on his hunts. They were sharp enough to easily break bone, but they couldn't cut completely through a rattler's meat and skin. To remove a rattler's head, he stood on its head and made a cut a few inches behind it. Then he'd grab the body of the snake and pull on it until the cut gave way, leaving the severed head underfoot. To remove rattles his technique was a little different. He made an initial cut on the tail an inch or two in front of the rattle. Then he stepped on the body of the snake and kept pulling and twisting his brush cutters until the tail and rattle came off. He never had any close calls with snakes that were still barely alive when he dismembered them in this manner.

A lady in his town who never wanted to see him pull into her driveway during the months of May through October was Alta Brown, the town clerk. She knew the possibility of his having a bucket full of rattlesnake heads and rattles was highly probable, and, according to Bill, she didn't want to see them or deal with them. Alta was a quiet farm woman whose father had been the town clerk for many years before she took over the reigns. She knew all the bounty hunters, but nobody brought in the volume of snakes that Bill Galick did. Bill told me he put forty to fifty timber rattlesnake remains in front of her on more than one occasion, and her records bear this out. He felt sorry for her because of all the paperwork it would create for her, so he used to give her tips. The procedure at the time was that she had to fill out a separate form for each snake killed. If a bounty hunter brought her ten snakes, she had to fill out ten forms. For every snake recorded, the state would pay ninety-five cents to the hunter and five cents to Alta for her efforts. That was how the one dollar bounty broke down. Bill realized that there had to be a better way to record a bounty hunter's snakes. He mulled the situation over for a while, then called Don Best, the town's representative, and explained all the paperwork Alta had to deal with because of his bounty hunting. Don brought the problem up at the state capitol in Montpelier, and from then on, no matter how many snakes a bounty hunter brought in, Alta only had to fill out one form, and she still presumably received a five-cent commission per snake.

Only one of the bounty hunters discussed in this book came close to making a living from the timber rattlesnakes he killed, and it wasn't Bill. It was Art Moore. In Bill's case, hunting timber rattlers was what he described as his summer job from 1946 through 1971. He didn't get rich doing it, but he made enough money to make it worthwhile. He told me on several occasions that he reported, or border-jumped, the bulk of his snakes to county clerks in New York

State during the 1950s and 1960s for five dollars apiece. If he killed two thousand timber rattlers in his career, he might have made somewhere in the neighborhood of ten thousand dollars by border-jumping them to New York. For an individual living off the land, that would have been a fair amount of money in that era. Border-jumping was widely condoned in those days, and nobody in New York State today is likely to bring charges against any of the long-retired bounty hunters who may have been involved in this fraudulent activity. For one thing, it would have been considered a misdemeanor at the time and thus subject to a two-year statute of limitations.

A much bigger source of income for Bill came every spring when he regularly trapped five hundred to seven hundred muskrats from the marshes on his farm. When the fur market peaked in the late 1970s, he was selling his best rats, a term commonly used amongst muskrat trappers, for as much as eight dollars a pelt. How good was Bill as a trapper? Dave Hicks, a well-known fur dealer in Hartford, New York, provided some interesting insight on this question.[2] Dave, and his father before him, bought every muskrat pelt that Bill ever sold. "Bill Galick was one of the premier muskrat trappers on Lake Champlain," Dave asserted. "He was innovative and creative in the way he went about trapping, and the last of a dying breed. When he dies, there will be no one to replace him. He was from an era of trappers who knew the outdoors and were forerunners of the conservation movement in their own way."

Dave continued to pile on accolades and say that there isn't a man on earth he'd rather sit with around the woodstove at his business and reminisce about the past. He went on about Bill's living a lifestyle that other people only dream about. He was able to live entirely off his land in the not-so-distant past, the second half of the twentieth century, by trapping, raccoon hunting, and timber rattlesnake bounty hunting primarily. He was a modern day pioneer in a way. We were both in total accord on this. At one point Dave explained how Bill and other top trappers from his era were conservationists. They knew when to stop trapping muskrats in the spring even if the season wasn't legally over. As soon as they saw any bite marks in any of the muskrats they trapped, they'd shut down for the year. Biting is part of the mating behavior of muskrats, and continuing to trap after seeing evidence of bites not only reduced the price a trapper could get for his pelts, but more important, it also would hurt the following year's crop of muskrats. This form of conservation may have been selfishly motivated, but it was conservation nonetheless.

Over the years Bill trapped thousands of muskrats because of his hard work and experience as well as the type of floating traps, better known as floats, he developed and used. But this wasn't his only area of expertise as a trapper. In the winter he trapped foxes, minks, and occasional coyotes that had come south from Canada and the Adirondacks via Lake Champlain. Dave told me that Bill was good in all areas of trapping and that there wasn't any wild animal he couldn't trap if he put his mind to it.

Bill's biggest source of income from his land often came in the fall when he hunted raccoons for the last three weeks of November in the rugged hardwoods on the mountain behind his farm. Although he could have started "coon" hunting earlier, he always waited until about the tenth of November to begin. Years of experience and knowledge garnered in his youth had shown him that raccoons didn't have their prime fur until then. In the late 1970s, when raccoon pelts were bringing the highest prices of his career, Bill claimed that he sometimes had one-thousand-dollar nights in the woods. It happened on the occasions when he killed twenty raccoons big enough to bring fifty dollars each from Dave Hicks. It was his norm to shoot between eleven and twenty raccoons per hunt. He only shot big ones for which he'd get between forty-five and fifty dollars apiece. Even on his less successful nights he'd rake in three to four hundred dollars.

How hard was the sport? It was extremely hard. There are not many people who could do it night after night without total exhaustion eventually overcoming them. This, in a sense, is where the men were separated from the boys, and where the casual "coon" hunters were separated from those, such as Bill, who made a large part of their living from the sport. He would hunt night after night for a week or longer before taking a night off. When hunting, he'd spend the entire night in the woods, get in a few hours of rest the next day along with his various chores, and then head for the woods again.

Although a physically and mentally tough man, Bill was never a superman. How could he deal with the weight of eleven to twenty raccoons? The answer was simple. He never did deal with the weight. As soon as the "coons" he shot hit the ground, he skinned them in three to four minutes and then moved on. The raccoons he was taking were heavy enough so he couldn't carry two at a time, he explained to me. He carried his pelts in back baskets and backpacks, and by morning he often had a heavy load just from his pelts alone. It wasn't until after a hearty breakfast that he'd take the time to flesh the fat off the skins.

I came to know many things about Bill Galick and his past. He was one of the most integrated individuals I have ever met. Everything about his life fitted together and made sense to me. When he arrived home from the Army after World War II, the take-home pay for mill workers in his area was a meager $38.00 a week. Bill could have committed himself to a life of drudgery in a mill, but it would have been a totally stifling experience considering his spirit of independence. More than anything else in life, he apparently wanted the freedom of living off his land. What's remarkable about the man is the way in which he accomplished this. Admittedly he had a phenomenal piece of property, which was loaded with game, and a sizable colony of timber rattlers at his disposal, but there's more to it than that. He was a man who perfectly blended his livelihood with his abilities and personality traits to achieve what he wanted in life, and therein lay his success.

In 1989 Bill's long relationship with the marshes and woods he loved so much changed forever. He was in his early seventies and the only man on the farm who was making any income other than Social Security. He couldn't deal with the taxes on his land, which over the years had climbed to seven thousand dollars a year. What the farm desperately needed was somebody from the younger generation to step forward and take over the reigns, but no one emerged. The upshot of this was that the Galicks decided to sell their farm to The Nature Conservancy. Nothing too jarring took place afterward. Bill and his elderly uncles, Steve and Tony, whose names were also on the deed, were allowed to live on at the old homestead, and pay taxes on four acres rather than two thousand. It was a blessing from Bill's point of view to be able to live on at his farm on the land he loved so dearly, and that's where he lived until the day he passed away in his mid-eighties in 2005. There certainly were worse scenarios that could have unfolded.

Why was The Nature Conservancy, which owns land all over the world, interested in the Galick property? They bought it primarily to help protect one of the northernmost colonies of timber rattlesnakes in New England. They also were interested in protecting the property's five-lined skinks, which are rare lizards in northern climes. Another reason they bought it, according to Bill, was because it is probably the wildest property in the entire state of Vermont. By purchasing what many locals refer to as "Galicks," The Nature Conservancy not only bought a significant timber rattlesnake den but also a great deal of the transient and summer habitat used annually by the snakes from the den. How important was this purchase to the timber rattlesnakes of Vermont? It could very possibly save the species from sliding into the abyss of extinction. If the Galicks had sold the farm to a developer, there's no telling how it might have been dissected and modified. Such a sale could well have spelled doom for the rattlesnakes that have lived there so long. There are times, and this was certainly one of them, when man has to do what's right for nature. As much as I support the acquisition of the Galick property by The Nature Conservancy, in the future whenever I'm in that neck of the woods, I'll always remember the rugged farm family and the old bounty hunter who lived there.

CHAPTER 9 : THE WALKING MACHINE
Art Moore

It was Bill Galick who first told me about Art Moore, very matter of factly and with a slight tone of reverence in his voice. It seems that there was something akin to a pecking order among the timber rattlesnake bounty hunters, and every individual knew his place within the order. Bill Galick was a highly skilled timber rattlesnake bounty hunter in his own right. No one ever doubted this. Art Moore was clearly in a class of his own, however, because of the thousands of timber rattlers he had killed while bounty hunting the snakes during the 1950s, 1960s, and first four years of the 1970s in New York State. Getting to know Art Moore led me into the fascinating history of timber rattlesnake bounty hunting in northeastern New York.[1]

During my first of several visits with Art at his scenic camp on Lake Champlain north of Whitehall, New York, I spotted two massive whitetail buck head-mounts in a room off his kitchen. The antlers on them were bigger by far than any antlers I had ever seen on a Vermont buck. There were other impressive bucks in the room as well. It turned out that in his life Art had taken five bucks from the Adirondack Mountains that weighed close to three hundred pounds. Any hunter in the northeast would recognize this as an incredible feat. After talking for about ten minutes, Art suddenly turned to me and said, "I thought you came here to interview me about timber rattlesnakes and my bounty hunting. I hunted deer for meat. I think of myself as a timber rattlesnake hunter first and foremost." At that point it was obvious I was with a master snake hunter and that the time had come to start finding out what it was about Art that made him a master.

A person doesn't arrive at Art's level of skill by chance. Everything about his or her past factors in. Art was born in 1932 in Washington County, New York, and raised on a small ten-acre farm at the southern end of Lake Champlain. He was so poor growing up he remembers his mother saying that the family only generated seventy-five dollars of income one year at the height of the depression. When he told me this, I interjected, "That probably wasn't enough to pay for the insurance on a car let alone buy one." "Car," he scoffed. "We didn't own a car." "How did you get to school?" I asked. "We walked," he replied. "How did you get provisions out to your farm?" I continued. "We went into town with a horse and buggy," he snapped.

Many of my generation, who were born at the end of World War II, never experienced hardships such as these, but the picture Art painted was certainly common in Vermont and upstate New York in the 1930s. Rural people in both states at the time were very self-sufficient when it came to dealing with their food needs, and this was certainly the case with the Moores. They grew their own fruits and vegetables and had chickens, pigs, and cows. They also hunted and fished. There was good hunting and fishing all around them. To help ends meet they also dug ginseng and trapped. Not surprisingly, Art's background is similar in a number of ways to those of Lester Reed and Bill Galick. All of these men certainly knew hardship, self-reliance, and the ways of nature early in life.

Around age nineteen, Art realized he could make money as a timber rattlesnake bounty hunter. Because of the vastness of the area he lived in, he felt he needed a mentor to help get him started. He began to follow a known bounty hunter in his area by the name of Francis Wilbur. Art claimed that Frank Wilbur, as he called him, was the best timber rattlesnake bounty hunter who ever hunted the snakes in the state of New York. Coming from Art Moore, that's saying a lot. One day Art was following Frank in the woods flanking a rattlesnake den when Frank suddenly stepped out in front of him at close range and said, "I know what you're trying to do. Don't follow me. Learn about these snakes on your own!" No combination of words in the English language could have hit Art any harder. It was the verbal equivalent of being hit on the jaw by a heavyweight boxer. Mr. Wilbur had made his position perfectly clear and in so doing had instilled in his follower a strong desire to become an excellent bounty hunter on his own. Art already had all the skills he would need: sharp eyesight, the strength in his legs to carry him quickly in the mountains from one rattlesnake area to the next, endurance, woods savvy, hunting ability, and more. Three sentences from a man like Frank Wilbur was all the mentoring he ever needed.

Art became a walking machine. Early in his bounty hunting career he knew the only way he was going to rise to the top was through a colossal amount of walking, among other things. As long as his family could remember, there

FIGURE 7. Rattlesnake bounty hunter Art Moore handling a timber rattlesnake on a den in Washington County, New York. Photo taken in 1979 by William S. Brown.

were reports of people seeing and killing timber rattlesnakes on certain mountains in Warren and Washington Counties. The initial challenge for Art was to leg it out and find out exactly where the snakes lived. He had a pretty good idea of where to start his quest by following Frank Wilbur as much as he had. Almost immediately he independently discovered and began to hunt a few dens in Washington County. He was on his way.

By the 1960s he was hunting around twenty dens a year. His stamina and determination were incredible. He had to walk over ten miles a day to cover approximately half of his favorite area. It's a mountain range approximately nine miles in length with dens scattered from one end to nearly the other. In the spring and fall of the year when the rattlesnakes were congregated on their dens, Art would start at daybreak at the north end of the range, walk as far down to the other end as time allowed, stopping at every den he knew, and be back by dusk. His pattern was to shoot all the snakes he could at a den and in any areas nearby. Then he'd take his hatchet and chop off their rattles and about three inches of their tails as fast as he could, and walk on. Occasionally he would linger to skin out particular snakes if they had exceptional size or coloration.

Where many of the strongest outdoorsmen would be lame for days after such a hike, Art would be ready to go again the next day. Often he would hunt that one mountain range on Saturday, hunt all day elsewhere on Sunday, then go to work on Monday and walk all day again as a routine part of his job. In the 1960s he worked in pest control for the New York State Department of Conservation, which later became known as the Department of Environmental Conservation, or DEC. His job took him into the mountains of Warren and Washington counties, the Catskills, and the Southern Tier counties along the Pennsylvania border. Almost every day of the year he'd be walking in the woods somewhere, often in timber rattlesnake habitat. One of his major responsibilities was to mark on maps any concentrations of oak trees that were infested with gypsy moth egg clusters. This made it possible for the Department of Conservation to target these areas for aerial spraying the following year. Another responsibility he had was to identify white pine trees that were infected by blister rust disease, then dig up or spray any nearby gooseberry bushes, which are the host plants for the disease. In the winter months it was not uncommon for him to be deep in the woods on snowshoes all day.

His job not only turned him into an endurance walker, it helped him discover timber rattlesnake dens he wouldn't have found otherwise. Anywhere he located dens while working for the state, he would invariably come back later to hunt them. In the case of dens outside the bounty counties, he would come back later to capture live snakes for sale to collectors. Art had one contact in Long Island who would pay him ten dollars per snake for live rattlesnakes. In short, the man's job, his bounty hunting, and his live animal collecting were a perfect blend.

Anyone who has the legs to get up to a timber rattlesnake den on the side of a mountain could easily kill a few snakes if he or she were there at the right time. As mentioned previously, timber rattlesnakes on their dens do not tend to act aggressively toward intruders. For the most part, they either try to escape from humans or stay in a coiled position. They'll often rattle when frightened, but they're just as likely to remain silent and never be seen because of their ability to blend into their environment. A child armed with the right weapon could easily do them in. Killing timber rattlesnakes was the easy part of bounty hunting in New York. What made it hard was the challenge of locating and hunting a number of dens that most other bounty hunters were either unaware of, or too lazy to hunt. Owing to the ruggedness of the terrain, especially in Warren and Washington counties, bounty hunters had their work cut out for them.

Art Moore was the most formidable timber rattlesnake bounty hunter in New York, not only because he knew the location of twenty or more hard-to-access dens, but also because he had the stamina and drive to hunt these dens and their associated basking areas regularly. Other bounty hunters of his caliber simply never emerged, at least not between 1950 and 1973. With Warren, Washington, and Essex counties all offering a bounty of five dollars for a dead timber rattler by 1956, a lot of money could potentially be made by the right kind of individual. A number of would-be bounty hunters must have been tempted by this reality but realized that they didn't have what it took to meet the challenge. Art basically hunted without competition and was the only person to rake in a lot of money.

The way the bounty resolutions were written by the boards of supervisors in New York's three bounty counties, a hunter was eligible to receive a bounty payment for any rattlesnake he killed. It didn't matter whether the snake was a neonate and had only what is referred to as a button rattle, or whether it was a fully mature snake with a multi-segment rattle. If a bounty hunter killed a pregnant female in any one of the three bounty counties after 1956, and if it happened to have nineteen fully developed young snakes inside, a larger number than normal but not impossible, the hunter could have made one hundred dollars by killing the snake, then cutting her open and killing her litter too.

Art killed a lot of pregnant females in his career, not just because of the money it could translate into, but also because of their vulnerability. Before giving birth, female timber rattlers typically bask in open, rocky areas that I have referred to as basking knolls. Art was fully aware of this vulnerability and used it to his advantage by hunting hard in September when the gravid females in his hunting areas typically give birth. Although he probably killed the bulk of his snakes on dens in the spring and fall of the year, he had days throughout the course of his career when he killed close to a hundred snakes, including newborns, on basking knolls in the month of September. Often the young snakes were fully alive, just prior to birth, inside the pregnant females he shot, but at other times they were recently born and bunched up for protection around their mothers. In either case, the little one-foot-long reptiles were easy to dispatch by foot and by sticks, but had to be dealt with carefully since a bite from even a day-old timber rattler could put a man in the hospital.

By hunting on weekends, vacations, and various days off at the best times to hunt the snakes, Art had three-hundred-snake weeks on numerous occasions. After his hunts, he'd typically drive to the nearest town clerk, often with a backpack full of rattles and tails, fill in the necessary paperwork, then head for some liquid refreshment. Eventually he'd receive his checks, some of which were for as much as fifteen hundred dollars and higher, from the county treasurer's office. Younger readers may not realize how large a fifteen-hundred-dollar paycheck was in the 1950s and 1960s. Only the most elite professionals of the day were making that kind of money.

One of the reasons Art was so adept at finding and killing timber rattlesnakes was that he knew so much about their behavior. Early in his career, for instance, he realized that they must follow the scent trails of other timber rattlers to certain areas and even to specific rocks after leaving their dens in the spring. He didn't need radiotelemetry to figure this out, just countless hours of observation in the woods. Besides knowing the location of numerous dens, he knew where the snakes tended to bask, what they tended to eat, and the areas they'd be most likely to hunt.

Art's knowledge of timber rattlers didn't come purely from his field observations and bounty hunting. He had a real scientific curiosity and conducted numerous homespun experiments to learn as much as he could about them. Early in Art's career his wife, Marjorie, realized that her husband was a serious,

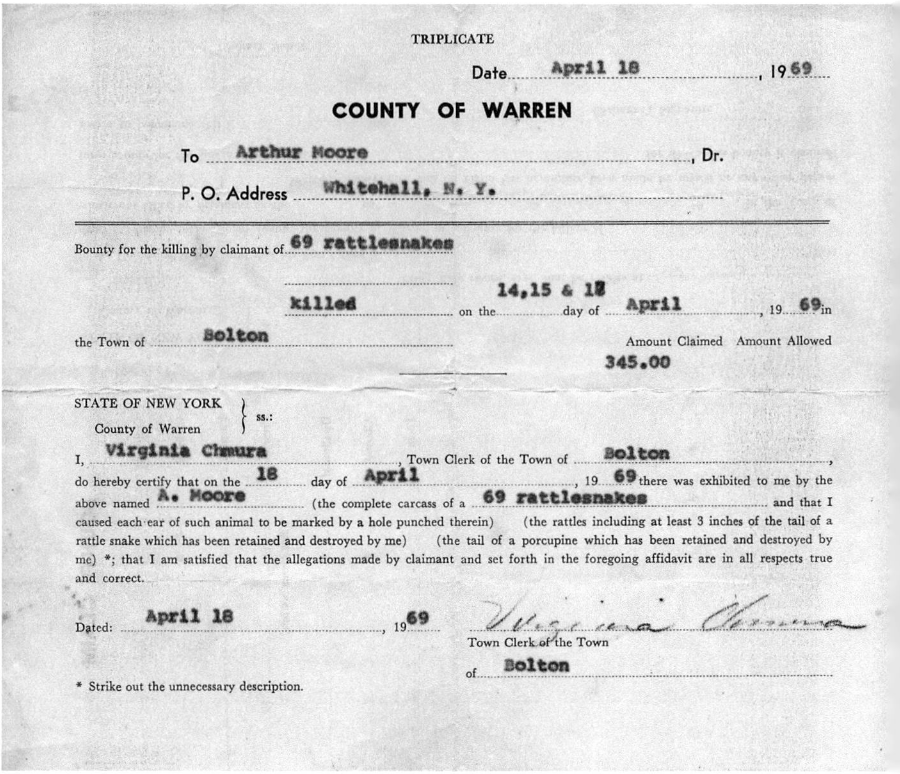

TRIPLICATE

Date...... **April 18** ...., 19 **69**

COUNTY OF WARREN

To ...... **Arthur Moore** ...., Dr.

P. O. Address ...... **Whitehall, N. Y.** ....

Bounty for the killing by claimant of...... **69 rattlesnakes** ....

.... **14,15 & 18**
killed .... on the ....day of ...... **April** ...., 19 **69** in

the Town of ...... **Bolton** .... | Amount Claimed | Amount Allowed

345.00

STATE OF NEW YORK ⎫
County of Warren ⎬ ss.:

I, ... **Virginia Chmura** ...., Town Clerk of the Town of ...... **Bolton** ....,
do hereby certify that on the ...... **18** .... day of ...... **April** ...., 19 **69** there was exhibited to me by the
above named ...... **A. Moore** .... (the complete carcass of a ...... **69 rattlesnakes** .... and that I
caused each ear of such animal to be marked by a hole punched therein) (the rattles including at least 3 inches of the tail of a
rattle snake which has been retained and destroyed by me) (the tail of a porcupine which has been retained and destroyed by
me) *; that I am satisfied that the allegations made by claimant and set forth in the foregoing affidavit are in all respects true
and correct.

Dated: ...... **April 18** ...., 19 ...... **69**

....
Town Clerk of the Town

of ...... **Bolton** ....

* Strike out the unnecessary description.

FIGURE 8. An April 1969 Warren County, New York, timber rattlesnake bounty claim form for sixty-nine snakes killed by bounty hunter Art Moore.

totally absorbed hunter and student of timber rattlers. When the Moores lived in Granville, New York, as a young married couple Art kept rattlers in cages in their basement and even under their bed. I'd like to know how many wives would stay with a man so driven. It's a testament to the love she obviously had for her husband. Keeping captive snakes continued for decades, and afforded Art the opportunity to make many interesting discoveries about his quarry.

The first area he wanted to understand better was hibernation. By exposing his snakes to cold temperatures both outside and even in the hydrator of his refrigerator, he was able to induce hibernation. At this same time he deliberately kept some of his snakes in constant warm conditions, and he observed that they never went into hibernation and remained active year-round. His friend and fellow timber rattler enthusiast in Long Island, John McDonough, conducted the same experiment of keeping timber rattlesnakes active year-round. Because of these experiments both men began to look at hibernation in a slightly different light and to think of it as a mode that the snakes drop into only if they have to.

Art wanted to know if there were other ways the snakes are able to regulate their metabolic rate to survive a serious situation. Normally he fed live mice or chipmunks to his snakes two to three times a summer, but he decided to see how long they could go without eating. So he let one rattler go for two years without food. All he gave it was water. The snake became so thin, according to Art's account, that he finally felt sorry for it and fed it six or seven mice over a period of several weeks. Miraculously, the tenacious reptile fattened back up and appeared to make a full recovery. From this he realized that the species is able to slow down its metabolism to deal with an acute shortage of food. When he described this experiment, I realized that he might have been a biologist or scientist in another field had he come from a different background, gone to school past the eleventh grade, and had the opportunity to go on to college.

Over the years Art almost always had rattlesnakes in cages outside his different residences. He would often let them out of their cages to see how they would behave and where they would go. If it was in the seventies and sunny, he noticed that they sometimes headed for shade but often lingered in the high grass areas near their cages. If it was sunny and the temperatures were in the mid-eighties or higher, they would instantly head for the nearest shade. By keeping timber rattlers in captivity, Art not only learned about their ability to regulate their metabolism, and their sensitivity to hot air, but also about their shedding, their biting of prey, and their pursuit of prey after biting it, among

other things. Anyone who came to know Art Moore realized that he had an in-
quisitive mind. It was a major asset to him in a number of ways in becoming a
bounty hunter of his caliber.

In one of my early interviews with Art at his camp, I asked him if he'd ever
been bitten by a timber rattler. He told me that he had had one serious bite
and one very frightening close call, which took place early in his career as he
was climbing up one of the steep ledges near his boyhood home. Just as his
eyes cleared a little shelf, a coiled rattler struck at his face. It missed him by
inches but shot venom into his right eye and onto his cheek. Art believed the
rattler was startled by him and struck defensively at his body heat. It was such
a close call that he nearly lost his footing and fell. Terrifically frightened by the
experience, he sat down and said he shook for ten minutes. The sting in his
eye was so bad afterward that he had to go to a local doctor for assistance. He
gave the distinct impression while describing this experience that the Moores
didn't make a habit of going to doctors unless it was absolutely necessary.

It's safe to assume that many beginning bounty hunters would have thrown
in the towel after such a fright, but for Art Moore, early retirement was out of
the question. He was already hooked on the challenge and excitement of rattlesnake hunting for life. It was a turning point in his career, however, since he
determined never again to be so careless. No matter how well armed and experienced he was at that time, he had clearly put himself in harm's way. He might
as well not even have been carrying his two pistols, a .38 Special and a .22 Special, both loaded with bird shot. It was pure luck that had saved him from being
struck and falling to his death that day, and he knew it.

In the summer of 1974 Art did the unthinkable, and alcohol may well have
played a part. He was careless once again with a timber rattlesnake, and this
time he almost died because of it. Early one evening after a long hot day, he and
two other men were enjoying a few beers and chatting at a popular local water-
ing hole. Before long the subject of timber rattlesnakes came up, and Art asked
the men if they'd like to see some rattlesnakes at his place, which was not
far away. Yes, they were definitely interested. Art was always willing to give
people a little education on his favorite reptile, so off they drove. Upon arriving
at his camp, he carefully removed a large adult rattlesnake from its cage. Mo-
ments later he was holding its head with the thumb and first two fingers of his
right hand and pinning its squirming, powerful body to his side with his right
elbow. Next, using his left hand, he gently passed a pencil through the snake's
mouth behind its fangs. It was a routine maneuver he had done on many occa-

sions. Right on cue the snake ejected two streams of venom through its fangs. The show couldn't have been going any better.

Just after Art finished "milking" the snake of its venom and lowering his left arm to his left side, one of the men yelled, "Watch out!" Art had not shut the lid on top of the cage, a couple of feet to his left, and a second big timber rattler was escaping. Instinctively he made a pushing motion with the palm of his left hand in the snake's direction. In a millisecond of poor judgment he must have thought he had a chance of blocking the determined reptile from coming out of its cage. He should have known better. The rattler nailed him with both fangs on his left index finger. Right away he knew he'd been bitten even though he hadn't seen the strike. Things happened quickly afterward.

First Art made sure both rattlesnakes were safely back in their cage, then he went inside his camp and told Marjorie that he'd been bitten by one of his snakes. She called their local doctor, who told her to have someone drive Art to the emergency room immediately. One of the men from the bar responded courageously to the crisis by sucking at least some of the venom out of the bite. Art had the presence of mind to first ask him if he had any cuts or sores in his mouth. The Moores' next-door neighbor rushed over and within minutes he and Art were on their way to the emergency room. About forty-five minutes later, as they were closing in on the hospital in Glens Falls, Art lost consciousness.

For three full days he was in a coma. The doctors surmised that such a severe reaction was largely triggered by the alcohol he'd imbibed prior to being bitten. It had somehow enhanced the toxic effects of the venom within his system. When he woke up, he was in critical condition in intensive care. After two more days in the ICU his condition stabilized to the point where he was sent to a regular room. Four days later he was finally released. A complication in Art's case was that he had such a reaction to the Wyeth skin test that his doctors decided not to give him any antivenom but rather monitor his conditions closely. Then at some point he developed a serious kidney problem as part of his reaction to the snake venom, so his doctors put him on a dialysis machine to save his life. It's not unknown for timber rattlesnake bite victims to develop life-threatening kidney complications in reaction to one of the elements in the snake's venom, but according to Dr. Keyler this is not a common occurrence.

As Art related his bite account to me, probably within a few feet of where his almost fatal bite had taken place nearly thirty years earlier, I began to wonder if he'd stopped dealing with timber rattlesnakes altogether after that experience. As it turned out, the bite may well have given him second thoughts about continuing his involvement with the snakes, but it certainly didn't stop him. At the time of the bite he was no longer bounty hunting, since Essex County had discontinued its bounty on timber rattlesnakes in 1971, and Warren and Washington counties had stopped paying bounties on the snakes the year before, in 1973, but he was collecting live timber rattlers, primarily for his high-paying customer from Long Island, and was not about to stop.

What may have driven him more than anything else to stay involved with the snakes was the excitement of dealing with them in a live state. It was one thing to find them and shoot them as soon as he saw them. It was a totally different experience to find them, pick them up, and put them into collecting bags. He found the potential danger involved in his new form of hunting more exciting than what he had routinely experienced as a bounty hunter. Another factor that drove him to stay involved with timber rattlesnakes after 1973 was his own reputation. He was widely known as a great bounty hunter and timber rattlesnake expert, and he had a reputation to live up to in his community. After his nearly fatal bite in the summer of 1974, he continued to capture timber rattlesnakes (as many as two thousand by his own admission) for sale to private collectors until 1983 when the state of New York added *Crotalus horridus* to its endangered species list as a threatened species. Then, and only then, did Art's long career as a rattlesnake hunter come to an end.

As intense a snake hunter and collector as Art was for over thirty years, there was a side to the man I'd be remiss in not describing. Every individual has a little yin and yang in his or her personality. Art's yang was his mischievous sense of humor and his genius for putting himself in comical situations. There probably wasn't a state trooper or local policeman in Art's area of New York State who didn't know his name during his heyday. What trooper wouldn't hear about a man who would go into barrooms and let a couple of rattlesnakes out on a pool table just to stir up a little excitement? At least twice Art cleared out barrooms this way and the police were called in. Once in Granville, New York, after he pulled this prank, one of the rattlers slithered down the corner pocket and had to be extricated with some degree of difficulty from the ball gutter!

On another occasion, two troopers pulled him over for some reason down near Binghamton, New York, where he'd been collecting live rattlesnakes all weekend and was on his way home with his stash of snakes. He had them in pillowcases, as was his norm, in the back of his station wagon, and all of the cases were bulging with the reptiles. Perhaps it was Art's attitude (he was prone to impatience) or his appearance, but somehow the troopers decided to push him a little. "What's in the pillowcases?" they asked. "Rattlesnakes," he replied. They didn't buy his answer and demanded to see for themselves.

Obligingly he got out, opened his tailgate, and with a trooper on either side of him, untied one of the pillowcases. Immediately several lively snakes got out and started moving toward freedom. Apparently the troopers almost jumped out of their skins! With a level of expertise clearly attributable to many years of experience, Art quickly corralled the snakes, and the open-mouthed policemen let him go. When Art described this experience, it was hard for him to suppress a chuckle or two even though he was unable to go to work that day because of back pain.

One time Art was coming down from a collecting trip in Essex County with three larger-than-average rattlesnakes in a pillowcase in the back of his station wagon when he decided to stop in for a couple of beers at one of the road houses he enjoyed. A man he knew was inside and wanted to see the snakes. As he went outside, Art warned him to be careful. Soon he came back in and said, "Art, you'd better come out." The snakes had somehow gotten out their case. One was coiled up by a gasoline pump. Another was under the station wagon, and they never did find the third snake, a fact that didn't seem to bother Art at all while driving home afterward, even though it might have been up under his seat for all he knew.

Art was definitely amused when he told me about an occasion when he came home one afternoon to find two men from the local telephone company working behind his camp. One man was up on a pole and the other man was feeding him wire from the ground. Around the base of the pole were several wooden boxes with plywood lids on them, which unbeknownst to the two workers were holding rattlesnakes. The man on the pole inquired what was in the boxes and Art told him rattlesnakes. He quickly slid aside one of the lids to show him. The man's phobia of snakes was so severe that he wouldn't come down until Art and the other worker moved the boxes about ten yards away.

In 2002 Art turned seventy. He was still as thin and fit as he was in his prime when he could walk all day in rugged terrain. He had great recall and enjoyed talking about his snake-hunting days. He was working several jobs including driving a garbage truck one day a week and operating a small tree-trimming service. As the animal control officer for his small town in New York, he was receiving calls almost every summer from people who had what is commonly referred to today as a "nuisance" timber rattlesnake on their property. The old bounty hunter loved to get these calls. They gave him an opportunity to once again deal with the snakes he had hunted so hard in the past, and the satisfaction of releasing timber rattlers back into the wild that not so long ago would typically have been killed by property owners in his area. In a sense, Art had gone full-cycle from a timber rattlesnake's worst nightmare to a person who would just as soon see them survive.

When historians look back at the twentieth century, it won't take them long to single out the 1960s as the most dynamic and event-filled decade in the United States. It was a politically turbulent era in which people began to question and challenge the status quo in a number of ways. One of the most significant outgrowths of the 1960s was the dawn of widespread environmental awareness, and one of the most important books of the decade was Rachael Carson's landmark *Silent Spring*, which blew the whistle on the dangers of the chemical DDT. After reading or becoming familiar with *Silent Spring*, many people began to worry about human poisoning of the environment, and about the health of the environment in general. Prior to Rachael Carson, there had been a handful of environmentalists in our culture, at best. The term environmentalist was seldom used. By the late 1960s a large segment of the American population was concerned about all the plants and fish and animals that in one way or another were being threatened by man's agricultural practices, habitat destruction, and different forms of pollution. At the federal level the groundwork was being laid for the national Endangered Species Act (ESA), which ultimately came into effect in 1973.

Vermont and New York were directly affected by the wave of environmental awareness that was sweeping the country by the early 1970s. In the first few years of the decade they established state endangered species lists, which led to state endangered species acts, patterned after the federal law. The two states also discontinued their bounties in those years. Alvin R. Breisch, one of the premier biologists with the DEC in New York since the 1980s, explained that New York's discontinuation of bounties on timber rattlesnakes in 1974, and its stopping bounties on all other animals (mainly bobcats, coyotes, and porcupines) in 1971, was directly linked to the new way of thinking about the environment and wildlife that arose in the 1960s.[1]

John Hall of Vermont's Fish and Wildlife Department described a similar scenario. When the state discontinued its bounty on bobcats and timber rattlesnakes, the only two animals still being bounty hunted in 1971, the powers-that-be in Montpelier felt it was no longer appropriate to have bounties on anything. They began to believe that it's more important to educate people about wildlife issues and realities than to try to eradicate or reduce a species. The Fish and Wildlife Department carries on its work in this same framework of thinking today, and John Hall acknowledged that the roots of this thinking clearly stem from the 1960s.[2]

By discontinuing their bounties on timber rattlesnakes, Vermont and New York took their first steps toward preventing the snakes from sliding into the abyss of extinction. Their second steps came in 1983 when New York added timber rattlesnakes to its endangered species list as a threatened species, and in 1987 when Vermont added them to its endangered species list. These steps constituted the first rays of hope for the conservation of the species in both states after over two hundred years of depletion at human hands. Once the snakes were thus protected, the stage was set for protecting their habitat in both states. Probably nothing is more important to the conservation of a species than protecting its habitat. Without its place to live, any species can face extinction.

The Galick property purchase by The Nature Conservancy in 1989 in Rutland County was, and remains to this day, the most important development in Vermont in terms of timber rattlesnake habitat protection. Shortly after the Galick purchase, the Conservancy took another important step in conserving the timber rattlesnake by acquiring the property on which the smallest of Rutland County's three rattlesnake dens occurs. Then, between 2000 and 2002, they began to buy a few key parcels of land near the county's second largest den. At this time they don't own the den, but they are beginning to protect critical summer habitat used by snakes from the den. In time they hope to acquire the farm on which the den occurs, but this may take a while to accomplish since the owners of the farm want to keep the farm in the family.

A number of important land acquisitions in New York by both the state and private parties have protected timber rattlesnake habitat in the former bounty hunting counties. Significant land acquisitions by the state occurred in 1981 and 1994 in Essex County. Both of these acquisitions have given protection to the den I have referred to as the northernmost den in northeastern North America, and have protected a vast amount of the habitat around it as well. A

Warren County acquisition that is pending while this book is in press will protect a sizable piece of timber rattlesnake habitat, including a major basking and gestating knoll, and will add it to a large tract of neighboring land already owned by the state. At one point in the recent past the two landowners were planning to build fifty-nine housing units and condominiums on the property. Had this taken place, it would almost certainly have had a detrimental effect on the large timber rattlesnake colony in the area. If New York is able to acquire this land it will be a significant victory in what could justifiably be considered an ongoing war to protect the species.

The state of New York is to be commended for the steps it has taken to protect its timber rattlesnakes; however, it can't be expected to buy every piece of timber rattlesnake habitat that becomes available. There's no question that the state has done its fair share, but private parties have begun to enter the process too, particularly in Washington County. I am aware of several private purchases of timber rattlesnake dens and important habitat near dens in Washington County that took place around the turn of the twenty-first century.[3] In all these cases the sole intent of the buyers was protection of the snakes, and thus their purchases seem to represent a new wave in which private parties are willing to take the protection of *Crotalus horridus* into their own hands in New York.

Two recent legal cases in New York have also had a major impact in protecting the state's timber rattlesnake habitat. With more cases likely to occur in the near future as man and the timber rattlesnake get closer and closer to each other in many areas, it's hard to imagine a case that will ever have a greater impact on preserving the habitat of this snake in New York than the Sour Mountain Case in 1999.[4] The preliminary injunction and temporary restraining order granted in that case by the New York Supreme Court, and later affirmed by the Appellate Court, will make it easier for the DEC to protect the habitat not only of the timber rattlesnake but also of any other endangered or threatened species in the future.

To understand this case, how it came about and what it could mean to future cases, we have to go back to 1990 when Sour Mountain Realty applied to the DEC for permits to construct an open-pit rock mine on a large property it owned in the southeastern part of the state. Initially, everything went fine with the permit process. Sour Mountain prepared a mandatory EIS (environmental impact statement) in which it concluded that any mining activities it might engage in on its property, such as blasting, would not have a significant effect on any timber rattlesnakes in the area. To fortify its position, Sour Mountain hired consultants who concluded that there were no timber rattler dens on or near the property. The DEC was apparently not too concerned at this point. There were no known timber rattlesnake dens on the property, and the nearest den of which they were aware was over a mile away. Late in the summer of 1996, while a hearing on Sour Mountain's EIS was in progress, the picture suddenly changed. The DEC received word that there was a strong likelihood of a previously unknown timber rattlesnake den being in close proximity to the Sour Mountain property.[5]

What transpired after this was an attempt by Sour Mountain Realty, on the one hand, to make sure that its large parcel of property would not be adversely affected by the proximity of some timber rattlesnakes, and an attempt by the DEC, on the other hand, to make sure that Sour Mountain would not threaten its neighboring timber rattlers in any way. In January 1997 the DEC was concerned enough about a possible new den to request a supplemental EIS (SEIS) from Sour Mountain Realty, which the mining company decided not to submit. Instead, it challenged the need for providing an SEIS through the Supreme Court in Ulster County, and lost.

During the next two years, to my knowledge, Sour Mountain never complied with the DEC's request for an additional environmental impact statement, even though Randy Stechert, in everyone's opinion, had located a new den not far beyond Sour Mountain's boundary line in the spring of 1997. Then, in January 1999, the realty company informed the DEC by letter that they intended to install a snake-proof fence along their property boundary near the recently discovered den. They indicated that the purpose of the fence was to block any timber rattlesnakes from the den from entering their property. The DEC wrote back a warning that, in effect, said that installing such a fence could constitute an illegal taking of the snakes from the nearby den under New York's ESA guidelines.

Sour Mountain next did something extremely unwise, in my opinion. They went ahead during the winter of 1999 and started to install the fence without any regard to the DEC's warning. To make matters worse, they didn't bother or attempt to acquire the necessary permits from the DEC to erect such a structure. They were already skating on thin ice with the DEC for not providing the supplemental EIS, which was requested two years earlier. The fence would prove to be their downfall. It's obvious that what they were trying to accomplish by installing the fence was to mitigate any negative effects that mining on their property might have on any of the timber rattlesnakes from the neighboring den. They seemed to be operating under the assumption that their snake fence would alleviate any concerns the DEC might have regarding the snakes. Installing the fence was nothing short of an act of futile desperation on their part that ended up being totally counter-productive, rather than helpful, to their long-range plans for their property. Here was a company that should have been trying to accommodate and keep on the good side of the DEC. At this point, with the confirmation of rattlesnakes being nearby, a little intelligent diplomacy could have done wonders for them. Had they been able to provide some, things might have turned out a lot differently than they did.

On March 1, 1999, the DEC went to the Sour Mountain site, confirmed that the company had begun to install a galvanized wire fence along its boundary line near the new den, or Sour Hill den as I'll refer to it, and determined that it had been installed in such a way as to block any but the smallest of rattlesnakes from passing through onto Sour Mountain's property. At this point the DEC contacted the Attorney General of New York and requested enforcement of the State's ESA against Sour Mountain Realty. (In New York, as previously mentioned, threatened species such as timber rattlesnakes are included in the state's endangered species list.) Within days the State of New York filed a complaint in the Supreme Court of Dutchess County alleging that the fence was in violation of New York's ESA. The State also moved for a preliminary injunction against Sour Mountain to remove the fencing it had already erected, and for a temporary restraining order to stop the defendant from installing any additional fencing.

On March 5, 1999, the court granted the State's application for a temporary restraining order, and a hearing regarding the preliminary injunction against Sour Mountain to remove the fence was held on March 19 and 22. The DEC came to the hearing well prepared to do battle. They had no intention of losing this case and were insightful enough to realize how important it could be as a precedent in future cases involving timber rattlesnake habitat. To win their case decisively, they knew they would first have to explain the annual cycle and habits of timber rattlesnakes, including emergence from their dens in the spring, migration shortly afterward to their summer habitats for foraging, mating, and gestating activities, and returning to their dens in the fall for hibernation. They would also have to present compelling evidence that the snake fence would adversely affect the normal annual movement of the snakes from the Sour Hill den and that this would clearly constitute a changing or modification of the snakes' habitat. Furthermore, they realized that Sour Mountain would try to dispute or downplay the importance of much of what the State would claim.

The DEC provided two expert witnesses in Dr. Theodore Kerpez, a Senior Wildlife Biologist with the DEC, and Dr. William S. Brown. The witnesses did everything the State hoped they would do, and according to Al Breisch of the DEC, Brown's testimony was more than compelling; it was a tour de force.[6] As anticipated, Sour Mountain tried to dispute a lot of what the State had to say about timber rattlesnakes and how the fence could affect them adversely, but Brown's argument, based on his years of field experience, was too much for them to overcome. He presented clearly to the court a strong likelihood that some of the snakes from the Sour Hill den would try to pass through the snake fence and that it would impede their normal migration to foraging, mating, and gestating areas on or beyond the Sour Mountain property. In other words, the fence would prevent access to, or modify, their habitat. The end result of the hearing was that the court granted the preliminary injunction and ordered Sour Mountain to remove the fence by no later than April 3, 1999. Sour Mountain, as might well have been expected based on their previous pattern of dealing with the DEC, appealed the ruling, and the case ended up in the Appellate Court of New York over a year later, in the fall of 2000. The main basis for the appeal was whether the modification of a species' habitat could constitute an illegal taking of that species under New York's ESA guidelines. The Appellate Court found that the installation of the snake fence by Sour Mountain did indeed constitute a modification or taking of a protected species' habitat, unanimously affirming the Supreme Court's earlier preliminary injunction.

This was a monumental day for the timber rattlesnakes of New York. It was the first time in the state in which habitat, not just the species, came to the forefront in a court decision regarding an endangered or threatened species. The decision's major significance, in my opinion, is that it gave the DEC the experience of how to approach similar situations, should they occur, in the future. It was also a declaration, in effect, that New York intends to take its Endangered Species Act of 1972 seriously. I believe that the Sour Mountain Case was a time bomb waiting to happen, in a sense. Sour Mountain Realty is currently in a limbo; they still haven't obtained the necessary permits to mine their property and, by the look of things, probably never will.[7]

In the future when herpetologists and people who care about the decline of the timber rattlesnake in New York State look back at events that were critical in the species' stabilization and survival, two cases from the same period of time will share the limelight: Sour Mountain and Ramapo Energy.[8] The roots of the Ramapo Energy Case date back to 1999 when American National Power, the mother company of Ramapo Energy, applied to New York's Department of Environmental Conservation for a permit to construct an electrical generating plant on thirty-five acres of primary timber rattlesnake habitat in a scenic valley in the southeastern part of the state.

Right away, environmental groups banded together to fight this proposal on the grounds of a number of environmental issues, using the forum of well-attended public hearings to speak out against it. Environmentalists were concerned about the visual impact of the plant's proposed 180-foot-high smoke-stacks, air pollution the plant might produce, the permanent loss of thirty-five acres of land adjacent to Harriman State Park, the protection of the Ramapo watershed, and the threat to the area's timber rattlesnakes, which are protected under New York State law. During the next three years a battle between timber rattlesnake experts took place that finally culminated in September 2002 when Ramapo Energy suddenly withdrew its application and abandoned its plans to construct a power plant in the valley. As in the Sour Mountain Case, it was the DEC's determination to protect timber rattlesnake habitat that made the difference.

This was a highly publicized case, even though it never got to the formal hearing stage or to a governor-appointed Siting Board, the normal final stage in a power plant application process in the state of New York. The battle that took place was based on voluminous written testimony, which was available to and reviewed by all interested parties. The two combatants were Ramapo Energy with its lawyers and consultants on the one side, and the DEC-PIPC (Palisades Interstate Park Commission) team with its lawyers and consultants on the other. From the outset, Ramapo Energy was determined to build its plant. A number of things it did, the testimony it produced, and the mitigation attempts it offered made this abundantly clear. One of the first things the company did was to come in and chainsaw down thirty-five or so large trees on the knoll on which it wanted to build its plant. This was done in part to evaluate what the knoll might look like in a naked state with a large power plant and smoke stacks on it. Ramapo Energy may well have thought that once this clearing was done, it would be just a matter of time before its plant sat on the knoll and that there'd be no snags along the way. If this was the case, it was wishful thinking.

For years prior to Ramapo's application in 1999, many people had been aware of a large timber rattlesnake population and numerous active dens in the valley. (This is the valley where Charles Snyder, the Bronx Zoo employee I referred to earlier, was bitten and killed by a timber rattler in 1930.) By 1999 Randy Stechert had been monitoring several dens in the valley for over twenty years. So as soon as Ramapo Energy applied for a permit to build its plant, red lights flashed in the DEC's Endangered Species Unit. After the tree-cutting episode in the spring of 1999, a complicated chain of events took place over the next three years, culminating finally in a crushing defeat for the energy company.

Before 1999 nobody ever reported or documented seeing timber rattlesnakes on the would-be site of the future power plant. Then suddenly in June of that year someone working for an environmental consulting firm hired by Ramapo Energy discovered a rattlesnake on the proposed plant site and informed Randy Stechert, among others. Stechert was in the second year of a four-year rattlesnake den distribution and population survey in the valley at the time. Word of the rattlesnake discovery spread fast, and more rattlesnakes were discovered on the proposed building site knoll during the summer. Ramapo Energy argued that the company's tree cutting earlier in the year had attracted the snakes. The implication was that once you clear a knoll in the vicinity of timber rattlesnake dens, timber rattlesnakes will begin to use that knoll for basking purposes. The company team knew that establishing this point might ultimately be helpful in obtaining a permit. It might soften any concerns the opposition had about the energy company building its plant on a likely timber rattlesnake basking knoll.

The DEC-PIPC team, led by W. S. Brown, Alvin R. Breisch, Geoffrey Hammerson, Edwin McGowan, and Ted Kerpez, jumped on this fallacious argument and stifled it in a hurry. They pointed out that no biologist had yet made a survey or analysis of basking areas in the valley and that no one had any significant data to go by on timber rattlesnake habitat use in the area. In effect, they maintained that it was pure conjecture, at best, on the part of Ramapo Energy to claim that its tree cutting had attracted the snakes to the knoll when the knoll could well have been in use as a primary basking area for thousands of years, for all anyone knew. It certainly was in a prime location for activities such as basking and mating and shedding.

In 1999, while in the first year of a three-year radio tracking study for the developer of an adjacent property, Randy Stechert made some extremely important discoveries about the knoll where Ramapo Energy wanted to build its plant. First, he and his associates observed six timber rattlers there that summer, two of which mated. He implanted two others with small radio transmitters. Then, a few months later, the implanted snakes, both of which were large males, did something that gave Randy some real insight into the importance of the would-be plant site. Instead of migrating back to the same den in October, they went to two different dens (there are eight DEC-documented, active, timber rattlesnake dens spread throughout the valley). When basking knolls are used by snakes from two or more distant dens, they are considered by timber rattlesnake experts to be significant knolls.

In the following summer of 2000, nobody found a single rattlesnake on the proposed plant site. The DEC-PIPC team attributed this to too many people tramping around on it during the previous summer, a phenomenon widely recognized by timber rattlesnake researchers and surveyors with substantial experience in the field as the "spook factor."[9] Men such as Randy Stechert, who coined the term, and W. S. Brown learned early in their careers that the more disturbance they create in a basking area or at a den in a given year, the less likely they are to see the same volume of timber rattlesnakes in the same area in the following year. Individual snakes, if not the species in general, seem to develop a long-lasting memory for areas in which they have been disturbed or spooked. Not surprisingly, the Ramapo Energy team offered a totally different explanation for why the plant site was apparently devoid of rattlesnakes in the summer of 2000.

The company team knew that if the opposition could establish that human interference can cause timber rattlesnakes to stop using areas they normally frequent, it would hurt the company's chances of obtaining a permit to build a power plant on their chosen site. Ramapo Energy, realizing the seriousness of the situation it was in owing to the expertise and intensity of its opposition, hired an additional consultant to its team at this time. The new consultant flatly denied that heavy human disturbance in the summer of 1999 was the reason that no timber rattlers were found on the knoll in question during the summer of 2000. Instead he and his team's other experts argued that annual variations in the species' summer habitats are normal and to be expected.

While it is true that annual variations are known to occur in the summer habitats of timber rattlesnakes, they still do not explain away the "spook factor." The energy company's argument showed the DEC-PIPC team two things. First, it confirmed the extent to which Ramapo would go to get its permit. This was an even more desperate stand, in a way, than its argument that tree clearing on the right kind of area could bring timber rattlers on the run. Secondly, it showed the DEC-PIPC team how limited their opposition was in its knowledge of the behavior of northern, denning timber rattlesnakes, their use of summer habitats, and their memory of disturbances within their habitats. This was a turning point in the case. The DEC-PIPC team realized afterward that they could eventually overwhelm the opposition with their tremendous field knowledge and determination.

By 2000, Ramapo Energy had clearly lost some of its early important arguments in the case. It was time to attempt to mitigate the situation it now found itself in, or possibly face throwing away its chance of obtaining its building permit altogether. For mitigation purposes the company offered to take nature into its own hands, in effect, and create a basking area on an outcrop knoll next to its chosen plant site. The plan was to pile up thin rock slabs in several locations on the knoll that would, the company claimed, attract timber rattlers. This seemed like such a simple solution, but would it actually work?

The DEC-PIPC team attacked this proposed mitigation on several fronts. First, they argued that nothing like this had ever been attempted anywhere else. Therefore, there were no data to evaluate. They also claimed that the way Ramapo Energy proposed to pile up the rock slabs would create an unhealthy or stressful situation for any rattlesnakes entering into them, since the cavities within them would be too hot and too dry. Their best and clinching argument, in my opinion, against the artificial basking area stemmed largely from Randy Stechert's three-year radio tracking study for Ramapo's immediate neighbor. In the summers of 1999, 2000, and 2001, neither Randy nor anybody assisting him with his study saw a single timber rattler on the proposed mitigation knoll despite numerous trips there. The DEC-PIPC team claimed in their testimony that there was no inclination for the snakes from the valley to use the knoll in the first place, and that any attempt to attract them to the knoll by placing a few poorly designed shelter rocks on it would be futile.

Ramapo Energy was smart enough to back away from its concept of creating a basking knoll at this time, but it wasn't ready to give up its cause just yet. In a second attempt at mitigation, the company offered to buy a piece of land about a mile away on the west side of a large ridge in the area, claiming in testimony that this new parcel, which was almost the same size as the main site, was good timber rattlesnake habitat and would replace any habitat lost in building the plant as planned. The DEC-PIPC team knew that the energy company's proposal had little or no merit and was quick to point out its weaknesses. First they made it clear that most dens, not only in that part of New York (the Ramapo Mountains and Hudson Highlands) but all throughout the northeast, tend to face in a southerly direction. Therefore, it was not likely that any rattlesnake dens would be in the immediate vicinity of the second proposed mitigation site since the site and most of the land surrounding it faced to the northwest. They also pointed out that the second mitigation area would not tend to attract snakes from at least half of the valley's known dens. This was not wild conjecture on their part. They knew this from Randy Stechert's four-year survey as well as his radio tracking study in the valley. If anything, while some rattlesnakes might migrate to the second mitigation site, the majority of snakes from the valley's known dens would travel in an opposite direction during their annual migrations to their summer habitats.

Once again the DEC-PIPC team had the expertise, the credibility, and the damaging testimony to quell Ramapo Energy's continued attempts to get what they wanted. The Ramapo Energy team had underestimated the formidable strength of their opposition. They also may have exercised poor judgment in hiring timber rattlesnake experts who, unlike the experts on the DEC-PIPC team, were totally unfamiliar with the geographical area in question and documented behavior like the "spook factor." Everything gradually fizzled out after this. Then, in the winter of 2002, there was a closed-door meeting of the two teams. Before this point, Ramapo Energy may not have believed that they were up against a brick wall, but during the meeting it must have become apparent that a bunch of timber rattlesnakes was going to defeat them. There's no doubt that the meeting was pivotal because nothing much else was heard from Ramapo Energy until they suddenly withdrew their application in early September.

Even though this case never reached a courtroom, it was an enormous victory, in my judgment, for the DEC-PIPC team, the timber rattlesnake, and all the people who banded together to fight the plant installation from the beginning. Unfortunately, all threats to this scenic valley, its native timber rattlesnakes, and their habitat are by no means a thing of the past. There are about 1,500 acres in the valley with an uncertain future. Currently a strong movement comprising various groups is seeking to persuade the state to purchase as much land as possible in the valley and annex it to the state park land it borders, but this could take a long time to transpire. In the meantime, it is likely that one business after another will come along with proposals to build in and therefore modify the valley's timber rattlesnake habitat. There is already a quarrying company on the horizon with plans that would involve a larger loss of critical timber rattlesnake habitat than that proposed by Ramapo Energy. I anticipate nothing short of a long, drawn-out war in the valley with many battles still to come. If so, it's clear from this case, and the Sour Mountain Case before it, that the DEC is fully committed to fighting the war to save the state's timber rattlesnakes from extinction, and is prepared to fight any critical habitat battles with every weapon at its disposal.

Although the Sour Mountain and Ramapo Energy cases took place in southern New York State, I believe that the impact of these cases will spill out into the entire state of New York as well as into other states in the northeast, and will offer shining rays of hope for the conservation and survival of the timber rattlesnake in the region in the future. Sour Mountain and Ramapo Energy were classic examples of a state government standing up for and defending timber rattlesnake habitat where it would otherwise have been seriously modified and threatened. In years to come, it's likely that lawyers knowledgeable in environmental law and cases involving threatened and endangered species will be familiar with both cases and, in one way or another, will use them to defend timber rattlesnake habitat and the habitat of other protected species in the northeast for as long as we continue to expand into and alter our landscape.

When evaluating everything that's been done to help the timber rattlesnake in Vermont and New York since the early 1970s, three individuals stand out in my mind. When it comes to Vermont, nobody has provided a greater ray of hope for the species than Mark DesMeules while he was working for the Vermont Chapter of The Nature Conservancy, both as the Coordinator of the Vermont Natural Heritage Program and as the Director of Science and Stewardship in the 1980s. It was DesMeules who spearheaded the Galick property purchase in Rutland County in 1989, thus insuring the protection of one of the largest and northernmost timber rattlesnake dens in New England way into the future.[10]

DesMeules's perseverance in bringing this major acquisition to completion was extraordinary, and it might never have happened if it hadn't been for a remarkable set of circumstances. In the early 1980s DesMeules was working for The Nature Conservancy out of its office in Montpelier, Vermont. In that era it was known by the Conservancy that timber rattlesnakes were in western Rutland County. That's not what first drew DesMeules to the Galick farm, however. What brought him there was Vermont's only lizard, the five-lined skink. The way he knew about the skinks being at Galicks' stemmed indirectly from a visit he made to the Museum of Comparative Zoology at Harvard University in 1983. It happened that the museum had two catalogued five-lined skink specimens from Vermont. DesMeules knew how rare they were in Vermont and was bowled over. He noted that the specimens were donated by Kinsman Lyon (the man Lester and Carol Reed took on a rattlesnake hunt in 1965). This led him to Lyon's wife, who was honored by the fact that someone was so interested in her husband's passionate, life-long interest in amphibians and reptiles. Through her, Mark was able to communicate briefly with Lyon, who was dying of cancer in a Boston hospital. After he passed away, Mrs. Lyon wanted Mark to have all of her husband's field notes, which he found to be extensive, informative, and extremely well written. Among other things, they led him to the Galick farm where Lyon had collected the skinks during a field trip with Bill Galick in the mid-1960s.

Once he knew where the skinks came from, he was determined to go to the farm. He was a little edgy about calling to arrange for a visit, however, because of rumors he had heard about the Galicks living an old-fashioned lifestyle and being uncomfortable with outsiders. So, the first time he went to Galicks', he snuck in by canoeing across Lake Champlain and hiking in with a friend.[11] During his initial reconnaissance on the property, DesMeules knew it was special and that the Conservancy might be interested in buying it at some point. Not long afterward he called Bill Galick, introduced himself, and arranged to make a proper visit. His friendship with the Galick family developed much the way mine did. It started a little slowly and guardedly the first afternoon but gradually gained momentum over time. In one of DesMeules's first visits to the farm, Bill Galick took him up to the rattlesnake den. It was during the trip to the den that a vivid realization came over him. As far as protecting the timber rattlesnakes of Vermont was concerned, Galicks' would be the perfect property for the Conservancy to buy because of its isolation and minimal threat of human disturbance.

Because of the snakes, the skinks, and the broad ecological diversity at the Galick farm, Mark DesMeules became determined to do everything possible to bring about an eventual transfer of the property to The Nature Conservancy. What he didn't realize was the amount of time and effort it would take to bring this to completion. It took seven years and numerous visits to convince the Galicks to sell the farm, but that wasn't the only thing that made the acquisition so arduous. DesMeules also had to convince the Conservancy that the property was important to obtain. Nowadays the acquisition is recognized and embraced, but during the 1980s the Conservancy didn't have the same strong feelings for the property that DesMeules had, at least initially.

One day in 1989 DesMeules and Bill Galick were canoeing out on Lake Champlain when Galick suddenly turned around and announced that the family was thinking about selling the farm. DesMeules couldn't have been happier, but his work was still not over. By this point he had the Conservancy interested in the property. His final two hurdles were to obtain enough financial backing to accomplish the initial phase of the purchase, and keep the Galick family from changing their minds. The way things played out, he readily obtained enough grant money to initiate the purchase, but a problem developed with the Galicks. They kept changing their minds about selling. DesMeules drove down from Montpelier a number of times fully expecting them to sign over the deed to their farm, only to find that they had become squeamish at the last minute and couldn't go through with it.

These last minute flip-flops were both frustrating and disappointing, but Mark DesMeules never gave up hope. Finally, during a trip to the farm in November 1989, the Galicks were ready to sell. The legal paperwork was signed and The Nature Conservancy became the new owners of the property. This was a great moment in conservation that would never have taken place without the vision and tenacity of Mark DesMeules. With the acquisition of the Galick farm, The Nature Conservancy made a decisive move not only to protect Vermont's five-lined skinks and timber rattlesnakes but also to preserve one of the wildest and most scenic properties in the state. On a picture-perfect day the following spring, the Governor of Vermont, Madeleine Kunin, attended an open house at the Galicks' and fell in love with the property. She was taken up to see the rattlesnake den and was so impressed by the vast panorama below her that she later referred to Galicks' as Eden.[12] I believe she hit the nail right on the head, and I'm sure Mark DesMeules feels this way as well.

No others have done more to help the timber rattlesnakes of New York State than Randy Stechert and W. S. Brown. Since 1983, the year timber rattlers were added to the state's endangered species list as a threatened species, both of these men have contributed significantly to the conservation of these snakes, and they are still going strong at the beginning of the new millennium. As long as they stay active in their careers, they will continue to be shining rays of hope for the continued conservation of the snakes, a cause to which they have totally dedicated themselves for most of their working lives.

Randy Stechert grew up in northern New Jersey. After being introduced to his first timber rattlesnake den in southeastern New York at the age of thirteen in 1965, he has been involved with timber rattlesnakes both as an amateur and professional ever since. In 1981 he was given a major opportunity to help the snakes he so admires when the DEC's Endangered Species Unit in New York hired him to conduct a statewide timber rattlesnake den distribution and population survey. The rest is history. Stechert has been employed annually under various contracts by the DEC since then and has become one of their top field biologists. His job was tailor-made for an individual with his amount of passion for the timber rattlesnake, along with a commitment to do whatever he can to keep the species from becoming extinct. As a den locator, monitor, and population survey specialist, he is involved with timber rattlesnakes across a huge multi-county area in the southern part of the state almost daily from the time the snakes first begin to appear on their dens in the spring until they can last be seen prior to going into hibernation in the fall.

During his career he has contributed to the betterment of the timber rattlesnake in New York in many ways, but it could be argued that his most important contribution came two years after being hired. There's no question, according to others with whom I have talked, that the information he provided the DEC in 1983—that timber rattlesnake populations in New York were as much as 75% lower than they were historically, and had been reduced by as much as 60% in the previous hundred years alone—had a lot to do with the department's decision to add the snakes to the state's endangered species list as a threatened species.

Since 1983 Stechert has continued to provide den distribution and population surveys of approximately forty-five dens annually to the DEC, which gives the department an idea of whether the populations at those dens are holding their own, increasing, or declining. This is critically important information for the department to have, and if it weren't for Stechert's dogged legwork in the field, including repeated visits to many of these dens, the DEC would only be able to guess the population status of individual dens throughout the state and would have no concrete estimates to go by whatsoever. All in all, Stechert is aware of no fewer than 187 dens in New York, of which he has discovered at least 130 on his own. When it comes to population status surveys and finding dens, nobody compares to him.

Randy Stechert also helps the timber rattlesnakes of New York through his consulting work, which is often referred to him by the DEC, but also by private consulting companies. This involves his going into timber rattlesnake habitat and writing reports that go beyond den distribution and population surveys. The significance of these reports can come into play when developers want to

FIGURE 9. Randy Stechert with a 54″ 5.88 lb. male yellow morph timber rattler in southern New York State. Photo taken on 8/3/1999 by Jesse Jaycox.

build on, or otherwise alter, pristine timber rattlesnake habitat, as was the case with the Ramapo Energy Company, which wanted to build a huge electrical generating plant on what turned out to be a significant timber rattlesnake basking knoll. The DEC knew about the knoll's significance from Stechert's state-required annual reports stemming from his three-year radio tracking study in the valley between 1999 and 2001. Even though this study was conducted for a private company, a neighbor of Ramapo Energy's, Stechert's annual reports on the study were available to the DEC.

In those reports he not only confirmed the knoll in question as a significant basking area but also identified and confirmed eleven other basking knolls in the valley. He also pin-pointed the valley's eight known dens, summer foraging and mate-searching habitat, and even the general migrating directions taken each year by many of the valley's rattlesnakes. These facts, coupled with the DEC's argument, had a lot to do with the ultimate squelching of Ramapo Energy's building plans. In this case the DEC needed unassailable data to challenge Ramapo's EIS application, and Randy Stechert certainly provided it. He's considered to be a timber rattlesnake habitat expert not only by the DEC in New York, but also by many independent sources. The way he became an expert was through years of observing, capturing, and marking the reptiles in their environment, and also by tracking them using radiotelemetry in the last decade. When timber rattlers leave their dens in the spring across the state of New York, Stechert has more knowledge about where they are likely to go than anyone else in the state.

Randy Stechert gets a lot of respect from his fellow herpetologists as well as others for everything he has done and continues to do to help the timber rattlesnakes of New York. He's not afraid to roll up his sleeves and confront head-on anyone who breaks the laws protecting his favorite reptile. As an example, he twice helped DEC environmental officers bust an individual, referred to previously, who was capturing timber rattlesnakes in New York and selling them to collectors long after the snakes were protected in 1983. Ironically, this is the very man who showed Stechert his first timber rattlesnake den in 1965. There's no point in my mentioning this individual's name. He's well enough known by law enforcement personnel throughout the state of New York, as it is, but it's important for the reader to realize that there are people in our society who represent a real threat to the timber rattlesnakes that are hanging on in New York, Vermont, and elsewhere.

Stechert is currently involved in what might well turn out to be the most important work of his career. In the mid-1990s he took a number of newborn timber rattlesnakes from a well populated den, and introduced them for several years into a severely depleted den in southeastern New York. After W. S. Brown clipped a permanent number in their belly scales, Stechert knew he would be able to recognize these snakes in following years, if they survived. As it turned out, a number of them not only made it through their first winters but are still living on at their new den nearly ten years later. This experiment opens up possibilities that were not known in New York or elsewhere prior to this. Now, thanks to Randy Stechert's study with these snakes, an interesting prospect looms on the horizon. It now looks as if man, for the first time, will be able to intervene to help restore dens that are in trouble, or possibly even dens that have been hunted to extinction. Since it is estimated by Stechert that 60% to 65% of New York's dens have suffered some form of depletion, this is great news to anyone who cares about the decline of these snakes.[13] It couldn't have come at a more important time for the species.

At this point in his career, Stechert could arguably be best described as the most important contributor to timber rattlesnake conservation in the state of New York. He has provided approximately 75% of the timber rattlesnake data that the state's three principal conservation agencies (the DEC, the New York Heritage Program and The Nature Conservancy) currently use and rely on.[14] There wouldn't be any focused or realistic effort to save these snakes from extinction in New York without him. Stechert might not accept the parallel, but I see him as one of the state's two most important foot soldiers in the ongoing war to conserve the species. A prime example of this is when he stands in front of unhappy and sometimes unfriendly developers and explains to them that their development plans appear to be in conflict with New York's ESA protection of timber rattlesnakes. Even before the DEC is alerted or involved, in some cases, Stechert faces these situations with no qualms whatsoever. He's a fighter.

The other important foot soldier in the war to save the timber rattlesnakes of New York is Randy Stechert's long-time colleague and friend, W. S. Brown. He's a fighter too. I'm sure that anybody observing and knowing Brown as a boy would have seen that he could be destined for a career in biology or herpetology, or both, by his early teens. In those years he was avidly collecting reptiles and amphibians on and near his grandfather's farm in southeastern Pennsylvania and keeping a half dozen or more snakes as specimens and pets at any given time. He was also keeping a number of different species of turtles. During this period, he and his friend William S. Birkhead designed their own system of recording specimen details on three-by-five index cards, delivered at least one successful talk to their science club at school, and went to meet and talk with Roger Conant, the well-known curator of reptiles at the Philadelphia Zoo.

Where Randy Stechert was already focused on timber rattlesnakes by age thirteen, Brown's focus on these snakes didn't occur until he was in his thirties. Becoming an assistant professor of biology at Skidmore College in Saratoga Springs, New York, proved to be pivotal in Brown's career, in that it put him in close proximity to the timber rattlesnakes in the eastern foothills of the Adirondacks. After taking his first field trip into the mountains of Warren County and finding his first timber rattlesnakes in the wild, there was no turning back. It was like a moment of revelation for him. Suddenly the path he'd been on since early childhood made perfect sense to him. He would devote himself completely to learning as much as he could about timber rattlesnakes and make his information known to the scientific community. It was 1978, only five years after the discontinuation of the timber rattlesnake bounties in Warren and Washington counties, and Brown's long-term study was officially underway.

In 1978 very little was known about the timber rattlesnake. When Laurence Klauber wrote his monograph on rattlesnakes in the 1950s, he relied heavily on Steve Harwig's input to write his section on timber rattlers. Harwig, who was not a biologist but rather an avid amateur herpetologist, had been observing timber rattlesnakes in the mountains of Pennsylvania for many years and was quite knowledgeable about the species to a point. He knew the basics: spring emergence, color morphs, den locations, basking areas, etc. There were a few critical biological facts, however, of which he was unaware. When W. S. Brown emerged in the late 1970s as an enthusiastic young scientist dedicated to learning as much as he could about these snakes, the stage was set for him to discover and unlock these facts, and that's exactly what he did during the next two decades.

By 1981 Brown had ascertained that the timber rattlesnake was perilously close to extinction in a significant portion of its range in northeastern New York. It seemed evident to him that many years of bounty hunting had had a major impact on the timber rattlesnakes of Warren, Washington, and Essex counties. Three well-known historic dens in the area had already been hunted to extinction. On top of this, there were a number of ongoing threats to the remaining timber rattlesnakes in those counties. These threats included the collecting of rattlesnakes for sale to collectors and the senseless killing of the snakes by people who feared them and thought of them as being far more dangerous animals than they are now known to be. In addition to these threats, Brown was very concerned as a result of two of the biological observations he had already made, which later proved to be ground-breaking facts in terms of our understanding of these snakes in the northern portions of their range. It clearly looked as if female timber rattlers in the eastern foothills of the Adirondacks were not reaching sexual maturity at an early age or producing regular litters afterward. Coupling everything together, Brown realized that something had to be done and done quickly to save the species.

It so happened that the DEC was considering adding timber rattlesnakes to its endangered species list as a threatened species at that time and asked Brown for his input. Al Breisch was working for the DEC's Endangered Species Unit in the early eighties. He remembers Brown's follow-up report in the fall of 1981 and the impact it made. Brown was so strong on his biological facts and made such a compelling case in favor of helping the snakes while there was still time, that the DEC had no other choice than to eventually list them as a threatened species, and thus give them a wide range of protection against being taken, possessed, or sold under New York State law. In Al Breisch's opinion, it was Brown's letter that made the difference.[15]

Brown's report to the DEC in 1981 was clearly a turning point in his career. It established him as a foot soldier in the war to save the timber rattlesnakes not only of northeastern New York but of the entire state from extinction. It also established him as a biological expert to whom other people will listen when it comes to these snakes. This fact has come into play over and over again. There aren't that many people who are willing to step forward to assist the timber rattlesnake. What makes Brown so powerful and unique as an advocate for the species is his combination of biological expertise and the ability to make himself heard.

After contributing so significantly to getting the timber rattlesnake legally protected in New York in 1983, Brown continued to unlock the mystery surrounding the species through his ongoing field study to the point where he is recognized as one of the top timber rattlesnake experts and researchers in America. He has written extensively about timber rattlesnakes in scientific journals and magazines, including *National Geographic*. His *Biology, Status, and Management of the Timber Rattlesnake* (Crotalus horridus): *A Guide for Conservation* is, to my knowledge, the first book ever written that is devoted to the conservation of a snake. Brown has also established the only long-term data base on permanently marked, released, and recaptured timber rattlesnakes in the northeast.[16] In addition to establishing the first "nuisance timber rattlesnake" program in New York, he has given countless seminars in an attempt to change the public mindset regarding these reptiles. Overall, his contribution to the timber rattlesnakes of the state of New York, as well as to the field of herpetology, has been nothing less than enormous.

When all is said and done, however, I believe that what W. S. Brown has done to protect the habitat of New York's timber rattlesnakes will be his major legacy. As previously mentioned, Brown's oral testimony in the Sour Mountain Case, the first case in which the state of New York made it clear that it would fight to defend the habitat of its endangered or threatened species, was extremely powerful. Shortly afterward, his fifty-page written testimony proved to be a turning point in the Ramapo Energy Case. Although Brown might deny it, he was the DEC's main weapon in both cases. In addition to contributing so significantly to those two cases, Brown has been involved in key acquisitions of timber rattlesnake habitat in northern New York, both privately and at the state level, and has helped form management programs for a number of state-owned properties on which either rattlesnake dens or basking knolls occur.

A prime example of Brown's involvement in state acquisitions of timber rattlesnake habitat is currently taking place in Warren County. This is the property I previously referred to on which a developer and two brothers were planning to build multiple housing units. What I haven't mentioned is the part that Brown is playing in this attempted acquisition. He has known from the outset that if he could persuade the state of New York to acquire this property, it would protect, in one tract of land, one major den and a series of basking knolls of varying degrees of importance that would probably be unparalleled in the entire northeast. When he explained to the DEC in the early 1990s the importance of acquiring this land and adding it to the adjacent forest preserve system, they accepted his reasoning immediately. The catch is that the state still has not been able to consummate a deal with the landowners. Apparently these owners are anything but easy to deal with, but Brown has never given up hope. Now, thanks to his initial recommendation to the DEC, and also to the efforts of a tenacious conservationist, Michael Carr, a great opportunity looms for the continuing, undisrupted study of the timber rattlesnake in one of its northernmost locations in the northeast.

It's possible that there will never be another timber rattlesnake champion like W. S. Brown in New York or anywhere else. Fortunately, he came along at a time when timber rattlesnakes were flirting with extinction, especially in the New York and Vermont counties in which they had been bounty hunted for over half a century. The bounty hunting period was over but there was still an ongoing threat to the snakes, as there remains today to some extent, mainly from people who kill them out of fear. After playing a major role in getting the timber rattlesnake protected in New York in 1983, it wasn't long afterward that Vermont protected its timber rattlers. No one would dispute that Brown may have had something to do, at least indirectly, with Vermont's decision to protect its timber rattlesnakes in 1987. Above and beyond these accomplishments, his effort to protect critical timber rattlesnake habitat throughout the state of New York has been exceptional. In short, because of Brown and a small but ever-growing group of people who identify with his cause, there will be some degree of future for the timber rattlesnakes of New York and Vermont, and especially for the snakes I have herein referred to as the border rattlers of Rutland, Warren, Washington, and Essex counties.

CONCLUSION

As a result of everything I have learned about the timber rattle-snakes of Vermont and New York, I have undergone a transformation in the way I feel about these animals. When I began my research I was neutral in my feelings, but as I came to know various facts about them—how passive they tend to be when not disturbed, the importance of their role in the food chains of the habitats they frequent, and the amount of encroachment and killing to which they have been subjected—I found myself sympathizing with them to the point where I hope for their continued survival. It's gratifying to me that the timber rattlesnakes in the four former bounty hunting counties I have chronicled escaped extinction and are now in a position to make a gradual comeback. It's encouraging to see the private acquisition of critical rattlesnake habitat in Vermont, the combination of private as well as state acquisition of similar critical habitat in New York, and the degree to which the state of New York has been willing to fight to protect its timber rattlesnake habitat. In a sense I have come to feel the same way about timber rattlesnakes as did Vermont's Fish and Wildlife Department and New York's Department of Environmental Conservation when they eliminated their subsidized killing of these snakes in the early 1970s. My attitude toward these beautiful and potentially dangerous reptiles, which have long held their place in the forests of America, is that we should understand them, respect them, and leave them alone. We also need to continue to protect them as much as possible. They are worth protecting for a number of reasons, not least of which is the fact that they are symbols of our rapidly vanishing wilderness. Additionally, when we protect the timber rattlesnake, it often translates out to protecting some of our wildest and most scenic areas.

All of this being said, what does the future realistically hold for the rattlesnakes of Vermont and New York? To answer this question we have to think of two separate groups of snakes. The rattlesnakes on the Vermont/New York border in the four former bounty hunting counties make up one of these groups. All of the remaining timber rattlesnakes in the state of New York make up the other. The future for Vermont's timber rattlesnakes looks promising. This is largely due to the fact that much of the habitat around the state's three remaining dens in Rutland County has been purchased by The Nature Conservancy. It is also due to the fact that much of the habitat in which Vermont's timber rattlesnakes live is extremely rugged, difficult to climb up into, and therefore, generally inaccessible to the public. In addition to having secured much of the critical rattlesnake habitat in Rutland County, The Nature Conservancy has taken steps to monitor the county's largest dens and to make the public more aware of basic timber rattlesnake biology through seminars and mailings to key property owners. Everything considered, Vermont's timber rattlers are about as well protected as this species can be, and there is no reason to think this won't be the case well into the future.

The future for the rattlesnakes in New York's three former timber rattlesnake bounty hunting counties looks almost as promising. Much of the critical habitat, including dens, basking knolls, and foraging areas, has been purchased both privately and also by the State of New York, often with the primary goal of protecting the snakes from becoming extinct. Something that bodes well for the rattlesnakes in those counties is the nature of the habitat itself. Much of it is in more rugged and remote mountains than Rutland County's timber rattlesnake habitat and is therefore much more difficult to access. One of the facts that doesn't bode so well for the timber rattlers of Warren County is that many of its snakes migrate into areas where they come in contact with fast-moving vehicles and summer residents. W. S. Brown instituted his "nuisance rattlesnake" program in the early 1980s to help prevent the unnecessary killing of migrating snakes by year-round residents and summer vacationers. As discussed, this program has had great success, but Brown realizes that many timber rattlesnakes will be killed in the county in the future if they happen to be on either the wrong roadway or wrong property at the wrong time. Everything considered, however, the snakes of Warren, Washington, and Essex counties are about as well protected as they can be, and they should be around for many years to come.

The future for the rest of New York's timber rattlesnakes, which are located primarily in the mountains of the Southern Tier, the Hudson Highlands, and the foothills of the Catskills, looks less promising. As discussed, a war is currently going on to protect critical timber rattlesnake habitat in the southern part of the state, and so far two major battles have been won by those who care about saving the habitat of these snakes. The Sour Mountain and Ramapo Energy cases in southeastern New York have made it loud and clear that New York's DEC intends to defend critical areas of timber rattlesnake habitat whenever a situation arises in the future in which it can do so. At the time of this writing, there is another major timber rattlesnake habitat case looming in southern New York State. Typically, it involves a developer with grandiose plans on the one side, the DEC on the other side, and some pristine timber rattlesnake habitat at high risk in the middle. If this case goes the way the Sour Mountain and Ramapo Energy cases have gone, it will check a potentially enormous amount of disturbance to a colony of timber rattlesnakes that has survived in the same area for thousands of years. There are already some good indications that the case will be hard fought and won by the DEC.

Although there are reasons to be optimistic about the future of New York's timber rattlesnakes in the southern portions of the state, there is also reason for concern. We have already seen in southeastern New York the reality of what can happen to timber rattlesnakes that have massive housing developments constructed right smack in the center of their habitat. In one case, a developer was given the "go ahead," in effect, by the DEC to build a sprawling mountainside development as long as he erected a snake fence to deter the timber rattlesnakes from several neighboring dens, approximately a third of a mile away, from entering the development. The hope by the DEC in this case was that the fence would do its job, that the rattlesnakes would adapt to no longer being able to travel into and through the development, and that a kind of peaceful coexistence would develop for the snakes as well as for their human neighbors.

The fact is, however, that the peaceful coexistence is not working out. Randy Stechert has recently completed a four-year radio tracking study for the New York Nature Conservancy and the DEC to verify the effectiveness of the snake fence in this case. What he has determined is that the fence is not doing its job. Timber rattlers are getting under it regularly. On one of my trips to this site with Stechert he captured a large male timber rattler, implanted with one of his transmitters, in a wood pile behind a house on the edge of the development. It was the second or third time he had captured the same snake in almost the same place. The snake clearly wanted to be in that immediate area, and it knew how to get under or around the fence in order to accomplish this feat. On another trip to the site, Stechert and I made a concerted effort to locate any places along the fence, which is thousands of yards long, where a rattlesnake could easily get underneath it, and within ten minutes we located a dozen or more areas.

In fairness to the installers who erected the fence, it isn't always possible to erect a snake-proof fence in rocky conditions, and there were a lot of rocky areas to deal with in this case. The installers did an adequate job in areas where they could dig down into soil and bury the fence's wire mesh six inches or so deep, but they simply folded the bottom of the mesh over in many of the uneven rocky areas they came to. The cost and time involved in removing these rocks and installing the fence correctly would have been prohibitive. The end result is that the fence is unable to keep out any timber rattlesnakes that are determined to get to the other side of it, and once they make it to the other side, there is a strong likelihood that many of them will be killed by people either living or working in the sprawling development.

Yes, the timber rattlesnake is fully protected by law in New York State, but that may not stop a construction worker in the development from pounding a timber rattlesnake to death with a shovel, or property owners from doing the same thing to a rattlesnake if they see one in their back yard. After killings such as these, nobody would be likely to call the local game warden or police, making it unlikely that any local authorities or the DEC would ever hear about them, or that any charges would ever be pressed. Stechert has already found one of his transmitter-implanted snakes, obviously stoned to death, in the development next to a house that was under construction at the time. Certainly there are some in the development who would refrain from killing any timber rattlesnake they encounter, and would call local law enforcement instead, but they may well be outnumbered by those who would kill one of the snakes in a confrontation rather than seek help. Overall Stechert is extremely pessimistic, as am I, about the longevity of timber rattlesnakes in this area. It won't happen overnight, and it may not happen in the next two or three generations, but eventually, in Stechert's opinion, residents of the development will refer to the rattlesnakes that used to be in the area.[1]

If I were limited to three words with which to pinpoint the timber rattlesnake's nemesis throughout most of its remaining pockets in southern New York State, they would be, "the almighty dollar." It's very clear that habitat loss from development in one form or another will be the greatest threat to the survival of these snakes in most of the areas in southern New York where they still remain. What we need to realize is the force that generates most development. It's man's economic greed. Timber rattlesnake habitat loss is not going to suddenly disappear as an issue in southern New York because there will always be developers who hope to make a fortune by building on, or in some way altering, critical timber rattlesnake habitat in the region.

In the worst case scenario, various conservation groups will eventually fail in their effort to protect the timber rattlesnake in many of the widely scattered pockets where it continues to survive in southern New York. Even though there have been some very encouraging habitat protection developments in New York State, there is no guarantee that much of the timber rattlesnake habitat in the southern part of the state won't eventually shrink away and be altered to the point where it can no longer sustain the snakes. The DEC is in what could well be the beginning stages of a war to protect the state's remaining timber rattlesnakes. It has done well in some of its early battles, but there are many more to come. Although the DEC appears to be ready to fight these battles, it's probably not realistic to assume that it will win all of them in the future. In any cases it doesn't win, it may have to compromise, and, in so doing, allow for some development in timber rattlesnake areas. Only time will tell how things will play out. Whether one has an optimistic or pessimistic view of the timber rattlesnake's future in Vermont and New York, it's encouraging to see that these states are leading the way in the northeast in defending the ever-shrinking habitat of these snakes.

APPENDIX A

Resolutions 77 (1971) and 107 (1972)
of the Warren County, New York,
Board of Supervisors

RESOLUTION 77

WHEREAS, on March 14, 1969 the Warren County Board of Supervisors adopted Resolution No. 65 providing for the payment of bounties for the destruction of wolves, coyotes, coy dogs, bobcats and timber rattlesnakes, be it

RESOLVED, that the County Treasurer be and he hereby is authorized to pay claims properly filed with him as hereinafter provided for the sum of $5.00 for each rattlesnake killed in the County of Warren, and that all bounties paid for wolves, coyotes, coy dogs and bobcats be eliminated, and be it further

RESOLVED, that such payments are to be made in accordance with the following regulations and requirements:

1. The only person eligible to make a claim for a bounty must be a bona fide resident of Warren County and will be required to submit proof of same.
2. That the Supervisor of each town is hereby authorized to appoint such person, or persons, in his respective town for the purpose of properly identifying each rattlesnake.
3. Where a bounty is claimed for a rattlesnake, at least three inches of the tail, with all rattles intact, shall also be exhibited to such person and this portion of each rattlesnake shall be retained and disposed of by him.
4. Upon receipt of the proper identification form, duly signed by the designated person, the town clerk in the town in which the rattlesnake was killed, shall execute, in triplicate, an affidavit for each bounty claim which shall state, in substance, the name and address of such claimant and the date and place where the snake was killed, and he shall retain one copy of such affidavit and certificate to be filed in his office. One copy thereof shall be mailed to the County Treasurer, and the third copy thereof shall be delivered to the claimant, and be it further

RESOLVED, that this resolution shall take effect on the first day of August, 1971. [*Author's note:* Resolution 77 indicates that the County Board of Supervisors in Warren County had what they perceived to be a timber rattlesnake problem in 1971.]

Resolution 107, which was adopted on June 16, 1972, allows us to see exactly what the board's concerns were regarding the presence of timber rattlesnakes in their county. The resolution is worded as follows:

RESOLUTION 107

WHEREAS, the New York State Legislature enacted a law, Section 206 of the Conservation Law, Prohibiting the State or any political subdivision from paying bounties on any wildlife effective July 1, 1971, and

WHEREAS, the law states that in the event a local health officer determines that any wildlife is a health hazard, bounties may be paid, and

WHEREAS, the hunting season when rattlesnakes may be killed is here, and

WHEREAS, the rattlesnake population has been held to its present level by the use of bounties for the past half century, and

WHEREAS, this method is far superior to a salaried hunting program, and

WHEREAS, the rattlesnake does not have any natural enemies and it does not fall into any other ecological category, and

WHEREAS, there is a direct effect of the rattlesnake population as it relates to real estate values and the tourist industry and any increase in the population of the rattlesnake would have a detrimental effect in both situations, and

WHEREAS, the Health Officers of the Town of Bolton, Hague, and Queensbury have declared rattlesnakes a health hazard in these three towns and have informed the Chairman of the Board of Supervisors of such in writing, now therefore, be it

RESOLVED, that the Warren County Board of Supervisors goes on record as reinstating bounties in the amount of Five Dollars ($5.00) per snake in Warren County, effective June 20, 1972, for snakes killed in the Towns of Bolton, Hague, and Queensbury, and be it further

RESOLVED, that the payment of bounties be limited to Warren County residents only and be it further

RESOLVED, that the Warren County Board of Supervisors retain the right to accept or reject any bounty payments at their own discretion.

APPENDIX B

Resolution 184 (1971)
of the Washington County, New York,
Board of Supervisors

WHEREAS, during the 1971 Legislative Session a bill was passed and signed by the Governor stating that it shall be unlawful for the State or any political subdivision thereof to pay bounties on any wild life and

WHEREAS, assessed valuation of $16,000,000.00 there exists a potential hazard (rattlesnakes) and

WHEREAS, thirty or forty years ago the Legislators in Washington County were aware that a hazard existed and

WHEREAS, they had the foresight to impose a bounty system to control snakes thus protecting their people and

WHEREAS, the summer population increases in these four towns therefore the danger increases and

WHEREAS, the snake has very few natural enemies and without the bounty system the number of the snakes will increase and

WHEREAS, in the event someone, especially a child, is bitten by a large snake and dies, the very important assessed valuation would decrease in these four towns

WHEREAS, the State of New York, Department of Health has gone on record as stating that no health hazard, as such exists in Washington County and

WHEREAS, it is the opinion of this Board of Supervisors that this law was hurriedly passed without proper investigation, Now, Therefore, be it

RESOLVED, that this Board of Supervisors goes on record requesting our State Legislators to introduce a bill authorizing municipalities to pay bounties for killing rattlesnakes.

APPENDIX C
Resolution 73 (1972)
of the Washington County, New York,
Board of Supervisors

WHEREAS, the New York State Legislature passed a Law, Section 206 of the Conservation Law, prohibiting the State or any political subdivision from paying bounties on any wildlife effective July 1, 1971, and

WHEREAS, the law states that in the event a local health officer determines that any wildlife is a health hazard, bounties may be paid.

WHEREAS, the hunting season when rattlesnakes may be killed is here.

WHEREAS, the rattlesnake population has been held to its present level by the use of bounties for the past half century, and

WHEREAS, the bounty system has been effective in other states such as Florida and Georgia, and

WHEREAS, this method is far superior to a salaried hunting program.

WHEREAS, the rattlesnake does not have any natural enemies and does not fall into any other ecological category, and

WHEREAS, there is a direct effect of rattlesnake population to possible real estate valuation and any increase in the population of rattlesnakes would reduce high value real estate which in turn would reduce Town and Country valuation, and

WHEREAS, the Health Officers in the Towns of Putnam, Dresden and Fort Ann have declared rattlesnakes a health hazard in these three towns and have informed the Chairman of the Board of such in writing and it is anticipated that the Town Health Officer in the Town of Whitehall who is presently on vacation, will do the same.

Now, Therefore, be it

RESOLVED, that the Washington County Board of Supervisors goes on record as reinstating bounties in the amount of Five Dollars ($5.00) per snake in

Washington County effective June 1, 1972, for snakes killed in the Towns of Putnam, Dresden, Fort Ann and Whitehall, and be it further

RESOLVED, that the payment of bounties be limited to Washington County residents only and

RESOLVED, the Washington County Board of Supervisors retain the right to accept or reject any bounty payments at their own discretion.

NOTES

1. GETTING TO KNOW THEM (pp. 1–37)

1. William S. Brown, *Biology, Status, and Management of the Timber Rattlesnake* (Crotalus horridus): *A Guide for Conservation* (Lawrence, Kansas: Society for the Study of Amphibians and Reptiles, 1993), 4–5.
2. William H. Martin, e-mail to author, December 12, 2004.
3. Laurence M. Klauber, *Rattlesnakes: Their Habits, Life Histories, and Influence on Mankind,* vol. 1 (Berkeley California: University of California Press, 1956), 232.
4. Klauber, *Rattlesnakes,* vol. 1, 217–31.
5. Klauber, *Rattlesnakes,* vol. 1, 231.
6. The vibrating of tails is sometimes referred to as "defensive mimicry," implying that other snakes mimic rattlesnakes by vibrating their tails when annoyed, or frightened, or to warn away a potential predator. The weakness in using this term is that snakes in other parts of the world where there are no rattlesnakes are also known to vibrate their tails under the same circumstances. See Klauber, *Rattlesnakes,* vol. 1, 232.
7. Martin, e-mail to author, February 7, 2005.
8. Brown, telephone interview with author, March 2, 2003.
9. For a good depiction of how William S. Brown marks and permanently records individual rattlesnakes, see Brown, "Hidden Life of the Timber Rattler," *National Geographic* 172 (1987): 136. The number that he clips into a timber rattlesnake's belly scales remains visible as a permanent scar even after countless sheddings. There are two ways other than Brown's belly scale clipping in which individual rattlesnakes can be positively identified over a period of time. One way is through PIT (passive integrated transponder) tagging. This is a technology that involves the insertion of a small "pill" with a bar code on it under a rattlesnake's skin. The drawback to PIT tagging is that the sensor, which is able to pick up and display a "pill's" bar code, has to be held very close to an implanted snake. The other way of identifying specific snakes over a period of time is through radiotelemetry. The great advantage of radiotelemetry, a technology in which a small radio transmitter is surgically implanted, is that it enables researchers to locate and home in on implanted snakes up to a mile away. The disadvantage of radiotelemetry is that the transmitters are known to malfunction, and their batteries only last about two years before they have to be replaced. Researchers and field biologists have been known to lose contact with snakes whose transmitter batteries have worn out prematurely. Randy Stechert told me in a telephone interview on October 6, 2004 that he once lost contact with a timber rattlesnake whose transmitter wore out while it was hibernating.

10. Brown, *A Guide for Conservation*, 18.

11. Martin details canebrake shedding rates in an unpublished manuscript, "Biology and Ecological Requirements of the Timber Rattlesnake (*Crotalus horridus*)."

12. Martin, e-mail to author, February 9, 2005.

13. Rulon W. Clark, telephone interview with author, February 15, 2005.

14. Klauber, *Rattlesnakes*, vol. 1, 368.

15. David Chiszar, "Predatory Behavior in Rattlesnakes: Envenomation, Chemoreception, and Discrimination," Biology of the Rattlesnakes Symposium, Loma Linda University, January 15–18, 2005.

16. Knowing how prolific Art Moore was as a killer of timber rattlesnakes, it's a little hard to imagine how he could have observed a timber rattler at the base of the same tree for several days without killing it. The answer is that encounters such as this more than likely occurred when he was working as a pest control technician for the state of New York in counties that didn't have a bounty on rattlesnakes.

17. For a discussion of a common way in which timber rattlesnakes are known to hunt at the base of deciduous as well as coniferous forest trees, see W. S. Brown and D. B. Greenberg, "Vertical-tree Ambush Posture in *Crotalus horridus*," *Herpetological Review* 23 (1992): 67.

18. Klauber, *Rattlesnakes*, vol. 1, 410.

19. Brown, telephone interview with author, March 17, 2005.

20. W. H. Martin told me in an e-mail on February 28, 2005 that basking timber rattlesnakes will eventually become too hot on calm days with bright sunshine and an air temperature of only 65° Fahrenheit if they stay in the sun too long.

21. Brown, "Hidden Life of the Timber Rattler," 133.

22. R. W. Clark, "Diet of the Timber Rattlesnake, *Crotalus horridus*," *Journal of Herpetology* 36 (2002): 497.

23. Klauber, *Rattlesnakes*, vol. 2, 1033–44.

24. Randy Stechert, telephone interview with author, February 15, 2005.

25. Klauber, *Rattlesnakes*, vol. 1, 306.

26. Klauber, *Rattlesnakes*, vol. 2, 1041.

27. Brown, e-mail to author, March 4, 2005.

28. Stechert, telephone interview with author, March 2, 2005.

29. Klauber, *Rattlesnakes*, vol. 2, 1050.

30. Stechert, telephone interview with author, March 2, 2005.

31. Daniel E. Keyler, e-mail to author, November 29, 2004.

32. While it is true that hibernating in dens is the norm for timber rattlesnakes in northern and central portions of their range, not all timber rattlesnakes den communally. This is often the case, for instance, in the southeasternmost areas of their range where canebrakes, as they are commonly referred to, are known to take shelter, often alone, in any suitable area they can find during cool and cold weather.

33. In one study it was shown that timber rattlesnakes maintained a mean internal body

temperature of 10.5° Celsius or 51° Fahrenheit over a nine-month period of time including hibernation. See Brown, "Overwintering Body Temperatures of Timber Rattlesnakes (*Crotalus horridus*) in Northeastern New York," *Journal of Herpetology* 16 (1982): 148. W. S. Brown has told me on several occasions that as long as timber rattlesnakes maintain a body temperature in the high 30° Fahrenheit range while hibernating in the coldest months of winter, they can easily survive.

34. Raymond L. Ditmars, *Snakes of the World* (New York: Macmillan, 1931), 5.

35. Bill Galick, interview with author at his farm, August 11, 2002.

36. Martin, telephone interview with author, 3 September 2002.

37. W. H. Martin, "Phenology of the Timber Rattlesnake (*Crotalus horridus*) in an Unglaciated Section of the Appalachian Mountains," in Jonathan A. Campbell and Edmund D. Brodie, Jr., editors, *Biology of the Pitvipers* (Tyler, Texas: Selva Press, 1992), 260.

38. Howard K. Reinert and Robert T. Zappalorti, "Field Observation of the Association of Adult and Neonatal Timber Rattlesnakes, *Crotalus horridus*, with Possible Evidence for Conspecific Trailing," *Copeia* (1988): 157.

39. Ditmars, *Snakes of the World*, 5.

40. For an excellent overview of the varying internal body temperatures and the thermoregulatory behavior of timber rattlesnakes, see William S. Brown et al., "Movements and Temperature Relationships of Timber Rattlesnakes (*Crotalus horridus*) in Northeastern New York," *Journal of Herpetology* 16 (1982): 151–61.

41. Jim Richards, telephone interview with author, December 23, 2003.

42. My knowledge of timber rattlesnakes frequenting certain islands in a lake in the southeastern Adirondacks stems from numerous casual conversations with W. S. Brown. Brown has not yet published his data on the phenomenon of timber rattlers swimming to and frequenting the same islands over a period of years.

43. Keyler, e-mail to author, March 29, 2005.

44. Thomas Burton, *Serpent-Handling Believers* (Knoxville, Tennessee: University of Tennessee Press, 1993), 31.

45. Leslie V. Boyer, "Snakebite: Not Just a Young Man's Disease," Biology of the Rattlesnakes Symposium, Loma Linda University, January 15–18, 2005.

46. Brown, telephone interview with author, August 20, 2002.

47. The details of the fight that Jed Merrow witnessed in northeastern New York were relayed to me initially in two separate telephone interviews with W. S. Brown: one on March 13, 2003, and the other on March 16, 2003. Also see Jed S. Merrow and Todd Aubertin, "*Crotalus horridus* Reproduction," *Herpetological Review* 36(2005):192. Another person who has witnessed and was extremely fortunate to film a combat dance between two male timber rattlesnakes in Connecticut is Sam Thompson, who showed his film during the Timber Rattlesnake Symposium at the Tanglewood Nature Center in Elmira, New York, on April 4, 2004.

48. Brown, telephone interview with author, January 30, 2005.

49. Kirk Setser, "Reproductive Traits of Female *Crotalus polystictus*," *Biology of the Rattle-*

snakes Symposium, Loma Linda University, January 15–18, 2005. Another area where female timber rattlesnakes are known to reproduce annually is in eastern Kansas. Two-year reproductive cycles are also known in the area. See Henry S. Fitch, *A Kansas Snake Community: Composition and Changes Over 50 Years* (Malabar, Florida: Krieger Publishing Co., 1999), 41.

50. Brown, telephone interview with author, March 18, 2003.

51. I owe much of my knowledge of female timber rattlesnake reproduction in the northern fringe of its range in the northeast, including delayed fertilization, gestation, and time of birthing, to a telephone interview I had with W. S. Brown on March 18, 2003. Randy Stechert also contributed to my understanding of this area in a telephone interview on February 10, 2004.

52. The bulk of my knowledge of timber rattlesnake longevity stems from a telephone interview with W. S. Brown on January 21, 2005. Also helpful was the paper he presented at the Biology of the Rattlesnakes Symposium at Loma Linda University in January 2005 entitled, "Long-term Ecology of *Crotalus horridus*: Dens, Survival, and Longevity." The paper was co-authored by Marc Kery and James E. Hines.

53. W. H. Martin, "Life History Constraints on the Timber Rattlesnake (*Crotalus horridus*) at Its Climatic Limits," in Gordon W. Schuett, Mats Höggren, Michael E. Douglas, and Harry W. Greene, editors, *Biology of the Vipers* (Eagle Mountain, Utah: Eagle Mountain Publishing, 2002), 285.

54. Henry S. Fitch, et al., "A Field Study of the Timber Rattlesnake in Leavenworth County, Kansas," *Journal of Kansas Herpetology* 11 (2004): 23.

55. Stechert, telephone interview with author, February 5, 2003.

2. HOW BIG ARE THEY? (pp. 38–42)

1. The skins of timber rattlesnakes, as is true with snakes in general, are most colorful and vibrant immediately after they have shed their previous skins. The big golden-colored timber rattler that impressed Bill Galick and his brother Ed so much was a yellow morph.

2. Art Moore, interview with author at his home, July 28, 2002.

3. William S. Brown, telephone interview with author, July 16, 2002.

4. For a good visualization of how W. S. Brown measures a rattlesnake in a squeeze box, see his article, "Hidden Life of the Timber Rattler," *National Geographic* 172 (1987): 136. Brown is measuring the tail of the timber rattlesnake in the picture as Randy Stechert gently holds down the rest of the snake's body against foam rubber with a piece of Plexiglas. Brown always measures the tail (that portion of a snake's anatomy from its vent or cloaca to its rattle) in addition to taking the overall measurement of a snake. Prior to measuring a tail, it's customary for him to measure the length of a timber rattlesnake while its entire body is contained within the box. He accomplishes this by drawing a line with an erasable felt-tip pen on the Plexiglas along the snake's dorsal midline from the tip of its snout down to the base of its

rattle. Afterward, when the snake is released, he carefully measures the line drawn on the Plexiglas with a tool called a map reader. In Brown's judgment, this is the best way to measure a rattlesnake objectively in captivity (i.e., without undue stretching). After a snake is measured in this manner, the line left by the felt-tip pen is easily erased, and the cleaned Plexiglas is ready to be used again in future measurements.

5. Stechert, telephone interview with author, March 4, 2003.

6. See Raymond L. Ditmars, "Serpents of the Eastern States," *Bulletin New York Zoological Society* 32 (1929): 111.

7. See Raymond L. Ditmars, *The Reptiles of North America* (Garden City, New York: Doubleday, 1946), 367.

8. In one of my early interviews with Bill Galick, he showed me the photograph, which I have used in figure 6, claiming emphatically that the two snakes he's holding are both six feet long. Shortly afterward I sent a blow-up of the photograph to W. S. Brown for his evaluation. Based on Mr. Galick being approximately five feet, eight inches tall, Brown estimated the live lengths of the two snakes to be fifty-two inches long and fifty-four inches long respectively. He sent copies of the photograph to Randy Stechert and W. H. Martin, and they both independently came up with very similar estimates. All three experts concurred that Mr. Galick's two timber rattlesnakes were uncommonly large male specimens for Vermont or northeastern New York, but that they were nowhere near five feet, or sixty inches, in length let alone six feet as claimed by Mr. Galick. In summary, this photograph is a classic example of rattlesnakes that have been stretched out about as far as they are able to stretch, including rattles in measurements, and adding on an extra twelve inches for good measure to make the snakes more impressive.

9. A prime example of how much a skin can stretch after being removed from a dead timber rattlesnake was described to me by Randy Stechert in a telephone interview on May 26, 2004. In 1997 Stechert measured a sixteen- or seventeen-year-old male timber rattler shortly after its death. The large snake, which had been reared in captivity by W. S. Brown, was exactly fifty-five inches long from the tip of its snout to the base of its rattle. Stechert's friend's son, Robert, skinned the snake, taking great care not to stretch its skin in any way. Once removed, the skin was sprinkled with borax and sandwiched between two long boards for drying. A month later the fully dried skin measured exactly sixty-two inches. In spite of Robert's care, it had stretched seven full inches during the skinning process. How much could a fresh rattlesnake skin be deliberately stretched? Randy estimates that by pulling on one, then tacking it down to a board for drying, a fresh skin from a large adult snake might easily be stretched eleven to thirteen inches. With that in mind, whenever I see a long timber rattlesnake skin mounted on a wall in the future, I'll know that whether the skin was intentionally stretched or not, it came from a considerably smaller snake.

10. There is no question that human beings are fascinated by long and giant snakes. For an excellent discussion of this fascination in North America, see John L. Behler, "The Great American Snake Hunt: On the Trail of Monster Serpents," *Animal Kingdom* 78 (1975): 21–26. John Behler, who was the Curator of the Department of Herpetology for the Wildlife Conservation Society (WCS), the parent organization that operates the world-famous Bronx Zoo in New York City, mailed me some interesting information regarding this fascination in November 2002. Around 1910 Teddy Roosevelt offered a reward of a thousand dollars through the New York Zoological Society to anyone who could produce a thirty-foot-long or longer snake for exhibition purposes at the Bronx Zoo. A lot of people aren't aware of this, according to Mr. Behler, but the reward remained in effect, growing significantly in size, until it was finally discontinued in the 1990s because a few critics strongly believed that it stimulated the hunting of snakes and other animals for sale to exotic pet collectors. The WCS, which superceded the New York Zoological Society in 1995, would probably love to have a thirty-foot anaconda or reticulated python on exhibition. On the other hand, they realize that these giant snakes are currently being hunted to extinction for food and the exotic leather trade. Enough is enough. The society has initiated a giant snake conservation program and believes the time has come to leave alone the greatly reduced number of giant snakes that remain in the wild.

3. WHEN A TIMBER RATTLER BITES (pp. 44–63)

1. Alan Tennant reports that dry bites, or punctures that are totally free of toxins, occur about 15% of the time in pitviper bites. See Alan Tennant, *Snakes of North America: Eastern and Central Regions*, rev. ed. (Houston, Texas: Lone Star Books, 2003), 20. Daniel E. Keyler reports that the incidence of dry bites is as high as 20% to 25% in the case of rattlesnakes in his 2005 unpublished paper, "Venomous Snakebite: Herpetologists and Herpetoculturists Pre-Hospital Considerations." He reports a similar percentage in Daniel E. Keyler, "Venomous Snakebites: Minnesota and Upper Mississippi River Valley 1982–2002," *Minnesota Herpetological Society Occasional Paper Number 7* (2005): 21.

2. See Findlay E. Russell, *Snake Venom Poisoning* (Philadelphia, Pennsylvania: J. B. Lippincott, 1980), Color Plate I.

3. For Russell's findings on the frequency of necrosis in North American rattlesnake bite victims, see Russell, *Snake Venom Poisoning*, 297. Dr. David L. Hardy, Sr., reports seeing necrosis in fewer than 10% of the rattlesnake bites he treats in Arizona. See David L. Hardy, Sr., "A Review of First Aid Measures for Pitviper Bite in North America with an Appraisal of Extractor Suction and Stun Gun Electroshock," in Jonathan A. Campbell and Edmund D. Brodie, Jr., editors, *Biology of the Pitvipers* (Tyler, Texas: Selva Press, 1992), 405.

4. Headache and nausea don't appear to be widely discussed symptoms in timber rattlesnake bite victims, but they are certainly not unknown or unreported. See Lau-

rence M. Klauber, *Rattlesnakes: Their Habitats, Life Histories, and Influence on Mankind,* vol. 2 (Berkeley: University of California Press, 1956), 837–38.

5. Bob Fritsch, telephone interview with author, March 6, 2004.

6. Keyler, telephone interview with author, June 8, 2004.

7. Dr. Keyler is not the only expert in snake venom poisoning who believes that the principal cause of death resulting from rattlesnake envenomations is the loss of circulating blood volume as a result of fluid shift. See Tennant, *Snakes of North America,* 20. Also see Russell, *Snake Venom Poisoning,* 209–10.

8. In studying the postmortem reports of twenty people who died from rattlesnake envenomations, Dr. Findlay Russell noted that the majority of the victims died between eighteen and thirty-two hours after being bitten. See Russell, *Snake Venom Poisoning,* 209.

9. Milan Fiske described his serum sickness in an article in *The Conservationist,* a magazine published by New York's Department of Environmental Conservation in 1981. See Milan Fiske, "A Personal Account of Being Bitten by a Rattler," *The Conservationist* (New York Department of Environmental Conservation) 36 (1981): 31.

10. For a discussion of the administration of antivenom to patients who experience sensitivity or allergic reactions to the serum, see Russell, *Snake Venom Poisoning,* 325–31.

11. The fact that rattlesnake bites are capable of causing instantaneous pain is advantageous to the snakes in that it can deter a potentially fatal attack from an enemy such as a man or a predator. See Klauber, *Rattlesnakes,* vol. 2, 828.

12. There are two explanations for the lingering problems that W. S. Brown is having with his left thumb pursuant to his bite in 2003. The first has to do with what is referred to as compartment pressure. His hand was severely swollen, and the pressure that built up inside his hand and thumb owing to fluid shift may have been sufficient to cause some permanent nerve damage to his thumb. The second explanation has to do with necrosis. Brown's infusion of antivenom didn't begin until nearly four hours after his bite. If it had been started within two hours, he probably would have avoided necrosis altogether. As it was, he developed significant postbite necrosis in his thumb, extensive enough to require two to three painful debridements of dead tissue a week for over two months. Having lost so much tissue in his thumb may well have produced some, or possibly all, of the nerve problems that he is experiencing. Brown is a classic example of the often used medical expression: time equals tissue.

13. Bob Fritsch, telephone interview with author, July 25, 2004.

14. Keyler, "Venomous Snakebite: Herpetologists and Herpetoculturists Pre-Hospital Considerations."

15. The "tough-it-outers" who forego going to a hospital after being bitten and envenomated by a timber rattlesnake because of financial concerns are correct in assuming that their treatment could be costly. The total bill for treating Brown's 2003

envenomation, including hospitalization and follow-up debridements, was approximately eighty-five thousand dollars.

16. Dr. Keyler makes the point in his 2005 unpublished report on pre-hospital considerations for venomous snakebite that many bite victims make a big mistake by waiting to see how bad their bites become before taking action. For whatever reason or reasons they delay, they may be taking a big risk. Why? An envenomation that starts off looking like a mild case can become an advanced envenomation after a few hours. My research has revealed that waiting before taking action could also be risky in the case of timber rattlesnakes bites in some parts of the country especially. Timber rattlesnakes as a rule throughout the bulk of their range have mainly hemotoxic components in their venom, but this is not always the case. There are timber rattlesnakes with what's been identified as Type A venom, which is largely neurotoxic, in Oklahoma, southern Arkansas, Louisiana, southeastern South Carolina, eastern Georgia, and northern Florida. See Robert Norris, "Venom Poisoning by North American Reptiles," in Jonathan A. Campbell and William W. Lamar, *The Venomous Reptiles of the Western Hemisphere*, vol. 2 (Ithaca, New York: Cornell University Press, 2004), 692. Being bitten by a timber rattlesnake in one of the Type A venom areas, and not seeking out help, could prove to be a fatal mistake.

17. Keyler, e-mail to author, November 29, 2004.

18. For Fred Stiles's account of his timber rattlesnake bite in the 1930s, see Fred Tracy Stiles, *Old Days–Old Ways: More History and Tales of the Adirondack Foothills* (Fort Edward, New York: Washington County Historical Society, 1984), 55.

19. Fortunately, Brown made the effort while he was still in the hospital to write remarkably detailed notes about his June 19, 1985, bite and post-bite experience. They allowed me to depict as accurately as possible a first-hand account of what I have described as a typical timber rattlesnake bite experience.

20. For Russell's opinion on fasciotomies, see Russell, *Snake Venom Poisoning*, 319 and 327.

21. For a description of antivenom treatment for moderate envenomations by North American rattlesnakes or cottonmouths, see Russell, *Snake Venom Poisoning*, 312.

22. In order to shed any possible new light on Bob Fritsch's near-death experience and discuss any post-bite complications he might be experiencing, I interviewed Fritsch extensively on the telephone, and exchanged more that two dozen e-mails with him between March 6 and September 9, 2004. For an additional discussion of this case, see Thomas Palmer, *Landscape with Reptile: Rattlesnakes in an Urban World* (New York: Ticknor and Fields, 1992), 46–52.

23. A thready (fluttery) and/or racing pulse can be a sign of a serious rattlesnake envenomation because it usually occurs concurrently with a decrease in blood pressure. Although Bob Fritsch's pulse was not racing, and his blood pressure was normal when the rescue team reached him, his thready pulse was an indicator of things to come, a lowering of his blood pressure and the possibility of going into shock. See Klauber, *Rattlesnakes*, vol. 2, 829.

24. Although monstrous thirst, as described by Bob Fritsch in his 1986 bite case, is not a symptom that is commonly mentioned in timber rattlesnake bite cases, it definitely is known to be associated with these envenomations. See Klauber, *Rattlesnakes*, vol. 2, 837.

25. It is not uncommon to administer twenty or more vials of Wyeth antivenom in extreme rattlesnake envenomation cases. See Russell, *Snake Venom Poisoning*, 312–13. In southern California, Dr. Roy Johnson has treated over 450 rattlesnake bite cases in his career. The rattlesnake involved in most if not all of these cases was the Southern Pacific, known scientifically as *Crotalus helleri*. Southern Pacifics are extremely toxic rattlesnakes. When I interviewed Dr. Johnson at the Biology of the Rattlesnake Symposium in January 2005, he told me that he used eighty-five vials of antivenom in one case involving a three-year-old girl who was bitten by a Southern Pacific and that it was his norm to use up to sixty vials of antivenom and more in cases in which people have been severely envenomated by these snakes. He is strongly of the opinion that the biggest mistake you can make as a physician treating a severe rattlesnake envenomation is not administering enough antivenom.

26. From what I have been able to ascertain, the removal of rings and any other constrictive items, such as bracelets, after being bitten by a rattlesnake on the hand or forearm, is a prudent and widely recognized first aid procedure today. Of the numerous experts on rattlesnake envenomation with whom I am familiar, almost everyone agrees on following this procedure when possible.

27. Keyler, telephone interview with author, August 9, 2006. Also see Hardy, *Biology of the Pitvipers*, 407. Also see Roger Conant and Joseph T. Collins, *A Field Guide to Reptiles and Amphibians: Eastern and Central North America*, third ed. (Boston, Massachusetts: Houghton Mifflin, 1991), 36.

28. There has clearly been a reversal in thinking in the last forty years about first aid treatment for pitviper envenomations. Nowhere is this more exemplified than in the writings of Roger Conant. In 1958 Conant recommended staying calm after being bitten by a pitviper, and removing rings or bracelets if the bite was on the arm, hand, or wrist. See Roger Conant, *A Field Guide to Reptiles and Amphibians* (Boston, Massachusetts: Houghton Mifflin, 1958), 30. That was sound advice at the time, and it is still sound advice in the beginning of the new millennium. He went on to recommend the application of a tourniquet a few inches above the bite. He stipulated that the tourniquet should not be too tight, and that it should be loosened every fifteen minutes to allow for swelling. After applying the tourniquet, he recommended making a cut no more than a quarter inch deep through each fang puncture with a sterilized razor blade or knife. Next came suctioning. His recommendation was to use two suction cups simultaneously if available, or to alternate one suction cup between the punctures. Pursuant to these steps, he recommended the application of a wet compress, followed up by going to a hospital. By 1991 Conant no longer recommended the use of a tourniquet, or cutting, or even the removal of rings and

bracelets after a pitviper envenomation. The only first aid procedure he recommended was the use of a Sawyer Extractor™ pump in an attempt to extract some of the venom. Other than this advice, his only recommendations were to stay calm and get to a hospital or medical facility as quickly as possible after first calling ahead to alert the attending physician to prepare for a snakebite emergency. See Conant and Collins, *A Field Guide to Reptiles and Amphibians*, 35–36. Dr. Findlay Russell is another person who exemplifies the reversal in thinking on the best first aid treatment for pitviper envenomations. In 1968 he co-authored a manual entitled *Poisonous Snakes of the World*, which was published by the Government Printing Office and included first aid measures. These measures were largely directed toward treating both viper and pitviper envenomations and included the use of a constriction band as well as both cutting and suction. See Russell, *Snake Venom Poisoning*, 265. By 1980 Russell had clearly changed his thinking regarding constriction bands, cutting, and suction. It would be difficult to claim that the first aid measures he now espoused were first aid measures at all. In his seminars in this period of time he advocated putting a bite victim at rest, giving the victim reassurance, immobilizing the affected part, watching for any untoward reactions, and transporting the victim to a medical facility as soon as possible. It seems he was saying in effect that there is no significant first aid that can be of use in treating pitviper envenomations. If so, Russell's thinking in 1980 overlapped with the thinking of other experts in the field of snake venom poisoning who became prominent later. See Russell, *Snake Venom Poisoning*, 264. Also see Hardy, *Biology of the Pitvipers*, 405.

29. For discussion of the risks created by cutting deeply into fang punctures, see Tennant, *Snakes of North America*, 20–22. Also see Conant and Collins, *A Field Guide to Reptiles and Amphibians*, 36.

30. Regarding the futility of attempting to suction rattlesnake venom from fang punctures, see Tennant, *Snakes of North America*, 22. Tennant makes the point that pitviper venom absorbs into the tissue surrounding fang punctures in a matter of seconds. Therefore, realistically speaking, there really is no time to stop this absorption.

4. THE SURVIVORS (pp. 65–69)

1. Laurence M. Klauber, *Rattlesnakes: Their Habits, Life Histories, and Influence on Mankind*, vol. 1 (Berkeley: University of California Press, 1956), 128.

2. W. H. Martin, e-mail to author, November 30, 2004.

3. J. Alan Holman, *Pleistocene Amphibians and Reptiles in North America* (New York: Oxford University Press, 1986), 106–108.

4. Martin, telephone interview with author, March 15, 2005.

5. Anne Marie Clark et al., "Phylogeography of the Timber Rattlesnake (*Crotalus horridus*) Based on mtDNA Sequences," *Journal of Herpetology* 37 (2003): 148.

6. Martin, e-mail to author, November 25, 2004.

7. Martin, e-mail to author, November 30, 2004.

8. Holman, *Pleistocene Amphibians and Reptiles*, 111–12.

9. Martin, e-mail to author, January 27, 2005.

10. Martin, e-mail to author, November 30, 2005.

11. John H. Bailey, "A Stratified Rock Shelter in Vermont," *Bulletin of the Champlain Valley Archaeological Society* (Fort Ticonderoga, New York) 1 (1940): 16.

12. Martin, e-mail to author, November 30, 2005.

13. Mark DesMeules, telephone interview with author, April 5, 2003. For further details on the history of the timber rattlesnake in Vermont, see Mark DesMeules, "Vermont's Timber Rattlesnake: Historic Distribution, Current Status and Conservation Outlook," in Thomas F. Tyning, editor, *Conservation of the Timber Rattlesnake in the Northeast* (Lincoln, Massachusetts: Massachusetts Audubon Society, 1992), 4–5.

14. Daniel E. Keyler, e-mail to author, November 29, 2004.

15. William S. Brown, *Biology, Status, and Management of the Timber Rattlesnake (Crotalus horridus): A Guide for Conservation* (Lawrence, Kansas: Society for the Study of Amphibians and Reptiles, 1993), 22.

16. See Richard Stechert, "Distribution and Population Status of *Crotalus horridus* in New York and Northern New Jersey," in T. F. Tyning, editor, *Conservation of the Timber Rattlesnake in the Northeast* (Lincoln, Massachusetts: Massachusetts Audubon Society, 1992), 1. For a good overview of the timber rattlesnake in Vermont, New York, and Canada, see W. H. Martin, "The Timber Rattlesnake in the Northeast: Its Range, Past and Present," *Bulletin of the New York Herpetological Society* 17 (1982): 15–20.

5. THE PRINCIPAL ENCROACHMENTS (pp. 70–82)

1. Laurence M. Klauber, *Rattlesnakes: Their Habits, Life History, and Influence on Mankind*, vol. 2 (Berkeley: University of California Press, 1956), 1166–67.

2. Klauber, *Rattlesnakes*, vol. 2, 1185–86.

3. Klauber, *Rattlesnakes*, vol. 2, 1166–67.

4. Joe Bruchac, telephone interview with author, January 7, 2005.

5. Klauber, *Rattlesnakes*, vol. 2, 1084–1187.

6. Alexander Henry, *Travels and Adventures in Canada and the Indian Territories Between the Years 1760 and 1776* (New York, 1809), 175.

7. Carl H. Ernst, *Venomous Reptiles of North America* (Washington: Smithsonian Institution Press, 1992), 72–73.

8. J. Bruchac, telephone interview with author, January 7, 2005.

9. Thomas Palmer, *Landscape with Reptile: Rattlesnakes in an Urban World* (New York: Ticknor and Fields, 1992), 55–56. Also see Albert Mathews, "Rattlesnake Colonel," *New England Quarterly* 10 (1937): 343.

10. Palmer, *Landscape with Reptile*, 54.

11. Robert Rooks (Vermont Fish and Wildlife Department), telephone interview with author, February 23, 2004.

12. Alvin R. Breisch (New York Department of Environment Conservation), telephone interview with author, February 24, 2004.

13. Erika M. Nowak, "Community Ecology of Nuisance Rattlesnakes," Biology of the Rattlesnakes Symposium, Loma Linda University, January 15–18, 2005.

14. Klauber, Rattlesnakes, vol. 2, 1165.

15. Klauber, Rattlesnakes, vol. 2, 1019.

16. Klauber, Rattlesnakes, vol. 2, 1020.

17. On August 24, 1883, The Brattleboro Reformer, a newspaper in Brattleboro, Vermont, referred to two local men killing three large timber rattlesnakes on Mine Mountain across the Connecticut River in New Hampshire, and subsequently rendering nine ounces of oil which they sold for one dollar an ounce. On September 16, 1886, the Reformer referred to a well-known rattlesnake hunter in the area who had killed eight snakes in one hunt a few years previously, and produced over fifteen dollars worth of oil from them.

18. John L. Behler, "The Great American Snake Hunt: On the Trail of Monster Serpents," Animal Kingdom 78 (1975): 22.

19. James Ellsworth DeKay, Reptiles and Amphibians (1842), 57. This is volume 1 of Zoology of New York, which is part 3 of a thirty-volume set under the title, Natural History of New York (New York: Appleton and Company, 1842–1894).

20. William S. Brown, Biology, Status, and Management of the Timber Rattlesnake (Crotalus horridus): A Guide for Conservation (Lawrence, Kansas: Society for the Study of Amphibians and Reptiles, 1993), 29.

21. Art Moore told me on several occasions during 2002 and 2003, the period of time in which I interviewed him extensively and came to know him well, that he never kept track of the number of timber rattlesnakes he captured for sale to various collectors but that it was somewhere in the vicinity of two thousand snakes.

22. We'll never know exactly how many timber rattlesnakes this individual captured in the northeast for sale to private collectors, but Randy Stechert, who knew this man well as a boy, and was introduced to his first timber rattlesnake den by him in the 1960s, has told me several times that this prolific snake hunter had captured and sold more than four thousand timber rattlers in the northeast by 1980.

23. Breisch, telephone interview with author, February 24, 2004.

24. William. S. Brown, Len Jones, and Randy Stechert, "A Case in Conservation: Notorious Poacher Convicted of Illegal Trafficking in Timber Rattlesnakes," Bulletin of the Chicago Herpetological Society 29 (1994): 75–78.

25. William E. Peterson III (a volunteer den monitor for New York's Department of Environmental Conservation), e-mail to author, June 19, 2005.

26. Brown, A Guide for Conservation, 29.

27. Zadac Thompson, History of Vermont: Natural, Civil and Statistical, in three parts (Burlington, Vermont: Chauncey Goodrich, 1842), 119.

28. W. H. Martin, "The Timber Rattlesnake in the Northeast: Its Range, Past and Present," *Bulletin of the New York Herpetological Society* 17 (1982): 16–17.

6. BOUNTY HUNTING (pp. 84–99)

1. Timber rattlesnake bounties were never common in the northeast or elsewhere in the overall range of the species. Certain towns in Massachusetts had bounties on timber rattlesnakes dating back in time to the early part of the eighteenth century. These were long ago discontinued. Certain counties in Iowa, Minnesota, and Pennsylvania had bounties on timber rattlesnakes as recently as the mid-twentieth century. See Laurence M. Klauber, *Rattlesnakes: Their Habits, Life Histories, and Influence on Mankind*, vol. 2 (Berkeley: University of California Press, 1956), 987.

2. Joe Bruchac, telephone interview with author, January 7, 2005.

3. William S. Brown estimates that the Snake Hill den by Saratoga Lake was hunted to the point of extinction by the time of the Civil War. He explained this to me in a telephone interview on June 24, 2005.

4. To be certain of the years in which New York's three timber rattlesnake bounty hunting counties first enacted these bounties, I enlisted the aid of three key individuals: (1) Judy Garrison, the Deputy Clerk of the Essex County Board of Supervisors (2) Sandy Huffer, the Deputy Clerk of the Washington County Board of Supervisors, and (3) Joan Parsons, the Deputy Clerk of the Warren County Board of Supervisors. Parsons and Garrison were able to provide me with accurate dates from Warren and Essex counties by reading back through old resolutions of their respective County Boards of Supervisors. In the case of Washington County, I went to the county seat in Fort Edward, New York, in January 2003 and read through years of old resolutions myself until I came upon the date of 1894 as the year in which the county first enacted its bounty on timber rattlesnakes. Sandy Huffer was helpful on that occasion in providing me with the necessary records.

5. The individuals who wrote the resolutions were totally unaware of timber rattlesnake biology. Timber rattlesnakes are born with what is known as a prebutton. Approximately ten days later, as noted in my text, they undergo their first shedding, at which time their prebuttons become their buttons. At the time of their next shedding, the following year, they acquire the first segment of their rattles. They gain a new rattle segment every time they shed thereafter. It's not common to find timber rattlesnakes in the wild with long, multi-segment rattles that still have their buttons attached. This is due to the fact that it's common for these snakes to break off their rattles and grow new ones without buttons.

6. The dates between 1986 and 1951 in which monetary changes were made to the timber rattlesnake bounty in Warren County were researched and provided to me by Joan Parsons, the Deputy Clerk of the Warren County Board of Supervisors, in the summer of 2002.

7. In 1971 when the state of New York put an end to bounty hunting throughout the entire state, Warren County was paying bounties on bobcats, coyotes, and timber rattlesnakes. Apparently feeling the need to continue its bounty on timber rattlesnakes only, the County Board of Supervisors wrote up and adopted unanimously Resolution 77 on June 18, 1971, and Resolution 107, adopted on June 16, 1972. For the wording of these resolutions see Appendix A.

8. Brown, telephone interview with author, April 3, 2004.

9. By examining available bounty payment records in the late 1970s, William Brown was able to discover that Art Moore had killed thousands of timber rattlesnakes in Warren County in the 1950s, '60s, and '70s.

10. Randy Stechert, telephone interview with author, July 2, 2005.

11. W. S. Brown made this discovery during his examination of old bounty payment records in the late 1970s. In 1981 he wrote a letter to the Department of Environmental Conservation (DEC) in New York recommending that the state of New York protect its timber rattlesnakes by listing them as an endangered species. In the letter he referred to Art Moore's total kill of 2,459 rattlesnakes in Warren County in the final seven years of the timber rattlesnake bounty hunting period as an example of how much damage a highly skilled hunter could inflict on the species over a period of time.

12. Brown, telephone interview with author, August 20, 2002.

13. In some of the Warren County timber rattlesnake bounty payment records that William Brown examined it's apparent that there was a nucleus of approximately two to three dozen hunters who made claims on timber rattlesnakes in Warren County in the years in which the county was paying a bounty of five dollars per snake.

14. Sandy Huffer, the Deputy Clerk of the Washington County Board of Supervisors, provided me with a copy of Resolution 26, which raised the bounty on timber rattlesnakes from fifty cents per snake to three dollars per snake in Washington County in 1948. She also provided me with a copy of Resolution 38, which raised the bounty to five dollars per snake in the county in 1956.

15. For the wording of Resolution 184, which was written on December 28, 1971, see Appendix B.

16. For the wording of Resolution 73, which was written on May 19, 1972, see Appendix C.

17. Brown, telephone interview with author, April 3, 2003.

18. Judy Garrison, the Deputy Clerk of the Essex County Board of Supervisors, researched the history of the timber rattlesnake bounty in Essex County and provided me with copies of every resolution the county ever passed pertaining to timber rattlesnakes. Although there were a number of resolutions between what was termed Act No. 1 in 1892 and Resolution 61 in 1956 that dealt with the bounty on timber rattlesnakes, none of them instituted any changes to the one dollar bounty on the snakes, which was established in 1892. Resolution 61 raised the bounty from one dollar to five dollars per snake.

19. After 1956 only three more resolutions were ever written in Essex County pertaining to timber rattlesnake bounties. They all occurred in 1969. Resolution 30, adopted on February 3, 1969, rescinded and cancelled all prior resolutions authorizing the payment of bounties for rattlesnakes, coyotes, and bobcats. Resolution 30, in itself, is a strong indication that the Essex County Board of Supervisors did not perceive that the county had a problem with coyotes, bobcats, or timber rattlesnakes in the late 1960s. Resolution 72, adopted on March 3, 1969, reinstated the three bounties. I believe that Resolution 72 was an attempt by the Essex County Board of Supervisors to stay in line with the thinking of Warren and Washington counties, both of which were paying bounty claims in 1969. Resolution 165, adopted on July 7, 1969, waived the residency requirements established in Resolution 72 so that individuals making bounty claims within the county no longer had to be residents of the county.

20. Art Moore, interview with author at his home, October 16, 2002.

21. To verify the commencement of a one dollar bounty on timber rattlesnakes in the State of Vermont in 1895, see Arthur F. Rice's letter to the editor, "Vermont's Rattlesnakes," *Forest and Stream* 44 (1895): 389.

22. Even though the smallest den in Rutland County is within easy walking distance of the largest den in the county, probably very few bounty hunters ever knew about it. This den wasn't officially recognized as a den until it was discovered by Mark DesMeules and a friend in the early 1980s. DesMeules explained this to me in a telephone interview on April 5, 2003.

23. I have seen and obtained copies of Vermont timber rattlesnake bounty payment records, spanning the years 1899 through 1967, from the town in Rutland County where the most timber rattlesnake claims were probably made. These records, which cover all but a few years of the entire timber rattlesnake bounty hunting period in Vermont, indicate that Bill and Ed Galick made more bounty claims on timber rattlesnakes than anyone else by far. Starting in 1947, the names associated with all bounty claims were eliminated.

24. The records referred to in note 23 show that forty-five different individuals made timber rattlesnake bounty claims on rattlesnakes that had presumably been killed at Rutland County's second largest den between 1899 and 1946. They also reveal that many of these hunters only made claims on a half dozen or fewer snakes in their lifetimes.

25. The two counties in which border-jumping was most likely were Warren and Washington. As mentioned previously, Essex County increased its bounty on timber rattlesnakes to five dollars per snake in 1956, but it's very unlikely that any Vermont border-jumpers attempted, through a New York accomplice, to turn in Vermont timber rattlesnakes in Essex County. Since there is only one known denning area in Essex County, clerks there would probably have been suspicious of any claims involving a lot of snakes.

26. It was Bill Galick who told me about border-jumping the bulk of his timber rattle-

snakes to New York State town clerks. Because of his honesty, I can definitively say that Vermont timber rattlesnakes were border-jumped across the Vermont border into New York state. Bill had two different accomplices, as he called them, who reported his snakes. After these men received their bounty payment checks, cashed them, and gave Bill the proceeds, Bill would in turn give them a little commission.

27. For additional documentation of this case, see W. S. Brown, L. Jones, and R. Stechert, "A Case in Herpetological Conservation: Notorious Poacher Convicted of Illegal Trafficking in Timber Rattlesnakes," *Bulletin of the Chicago Herpetological Society* 29 (1994): 74–79.

28. Pat Leclaire, telephone interview with author, July 3, 2005.

29. Timber rattlesnakes are metabolically able to survive freezing temperatures for short periods of time and recover fully with no ill effects. A great example of this fact was related to me by W. H. Martin. When Martin was in high school in northern Virginia, he woke up one morning in the winter and was alarmed to see that a young adult timber rattlesnake he was keeping in a cage in his bedroom closet had frozen to the point where it was stiff to the touch and showed no signs of movement whatsoever. The thermometer in the closet read a chilly 28° Fahrenheit, and Martin estimated that the snake had been exposed to 28° for several hours. He had no way of knowing whether it was frozen to its core, but it was certainly frozen externally. Before leaving for school, he put the cage in the living room to see if the snake could revive. When he came home in the afternoon, it had completely revived and was rattling at him. He kept the snake for another year and a half before releasing it in apparent good health. It's entirely possible that some of the frozen timber rattlers that were left on Marion Guerin's front porch in the summer of 1971 or 1972 could have recovered and been rattling when she stepped outside late in the afternoon.

7. A FAMILY TRADITION: LESTER AND CAROL REED (p. 116)

1. Of all the individuals who became an important part of my book, Lester and Carol Reed were not only the first people I interviewed and came to know well, they were the easiest to deal with. After my initial interview with them at their home in western Rutland County in the winter of 2002, I had several additional interviews with them in their home and as many as two to three dozen follow-up telephone interviews with them over the course of the next year. From the outset they were extremely interested in my book and the direction I was taking with it. They went out of their way, therefore, to open up to me about their days of bounty hunting and oil making, and they always seemed to make an attempt to be as accurate as possible about the scores of facts and details they discussed with me.

8. A MODERN DAY PIONEER: BILL GALICK (pp. 117–124)

1. Although driving out to the old Galick farm and meeting Bill Galick for the first time was a nervous experience for a number of reasons, it was an unavoidable hurdle to

cross if I wanted my discussion of Vermont timber rattlesnake bounty hunting to have any historical relevance. This being said, I often wonder how I was able to crack through Bill Galick's super tough and guarded exterior. I can't limit my success to the fact that I went out to visit him at his farm on numerous occasions, or to the twenty-five or more telephone interviews and conversations we had. There were other factors that must have come into play. Perhaps luck factored in somehow, or the fact that we shared a number of interests in the out-of-doors together. It may have been my timing, however, more than anything else that allowed me to penetrate his facade when I came into his life late in the winter of 2002. In hindsight, it seems as if Bill was at a point in his life where he wanted to tell the truth about his bounty hunting days, and also about the way he was able to live completely off his land without ever making so much as a penny working anywhere else, except for the years he spent as an Army gunner on supply ships in the South Pacific during World War II. Considering the fact that he didn't make his money by farming primarily but rather by trapping, raccoon hunting, and bounty hunting, he must have sensed that he had led a very unusual twentieth century lifestyle that was worth relating to others.

2. Dave Hicks, telephone interview with author, January 2, 2003.

9. THE WALKING MACHINE: ART MOORE (p. 128)

1. Of all the individuals who played a significant role in this book, Art Moore was the hardest person to get to know. I had a good interview with him when I first met him in the summer of 2002, but I could see that he had only begun the process of looking me over and judging me, and was holding back. It wasn't until summer turned into fall, and we had had several additional interviews at his camp and on the telephone, that he began to really open up to me. I don't believe he would ever have told me about some of the escapades that I have detailed in this chapter if he hadn't arrived at a point where he was comfortable with me. Needless to say, it was escapades such as these that had a lot to do with Art becoming known not just as an exceptional timber rattlesnake hunter but as a legend in his own time. One of the signs that Art was really opening up with me was that he told me about how his friends turned snakes in for him in Warren County toward the end of the bounty hunting era when only residents of the county could collect a bounty on timber rattlesnakes. There had to be something other than our frequent interviews that triggered Art into gradually opening up to me. If he somehow realized that his demise was near when I was interviewing him in the summer and fall of 2002, and the early winter of 2003 (he died in the summer of 2003), he may well have thought that his days of kidding and tall stories were over. Whatever it was that made him open up and become seemingly honest with me, it didn't take me long to realize that if I wanted to write about timber rattlesnake bounty hunting in the eastern foothills of the Adirondacks, I had come to the right man at the right time.

10. RAYS OF HOPE (pp. 141–161)

1. Alvin R. Breisch, telephone interview with author, March 26, 2003.

2. John Hall, telephone interview with author, March 24, 2003.

3. W. S. Brown, telephone interview with author, September 2, 2002.

4. For an understanding of the Sour Mountain Case, see Christopher A. Amato and Robert Rosenthal, "Endangered Species Protection in New York after State v. Sour Mountain Realty, Inc.," *New York University Environmental Law Journal* 10 (2001): 117–45. Amato and Rosenthal were the lawyers from the New York State Attorney General's office who represented the state of New York in the Sour Mountain Case.

5. In a telephone interview on March 13, 2004 Randy Stechert explained how the DEC got the impression right in the middle of an important hearing that a timber rattlesnake den might be near the Sour Mountain property. Everything started in August 1996 when Jesse Jaycox, a biologist with the DEC, wrote a letter to the DEC indicating that he strongly suspected the presence of timber rattlesnakes on the mountain where Sour Mountain Realty planned to do its mining. Jaycox based his suspicion largely on his reconnaissance on the mountain, plus reports of timber rattlesnakes being seen and killed there since his childhood in the area. Stechert, one of the DEC's top field biologists for many years at the time, urged him to go back up the mountain to the general vicinity of the Sour Mountain parcel to see if he could locate any rattlesnakes. Jaycox went up and found three rattlers, one of which appeared to be a postpartum female, or a female that had delivered her young. Not long afterward, Jesse and Randy went up together and found a postpartum female with nine neonates. She was in the same location that Jesse's postpartum female had been in a few days earlier. During this field trip Jaycox and Stechert also found a gravid female and one other rattlesnake, leading them to believe that a den was nearby. The following spring, on May 15, 1997, Stechert located nine timber rattlesnakes together on a ledge near Sour Mountain's boundary line and assumed that he had found a den site. He was under a contract from the DEC to locate a den in the area and was fairly certain he had succeeded in doing so. An interesting footnote regarding Stechert's new den is that everyone involved in the Sour Mountain Case assumed that he had found a new, previously unknown den. Locating dens is Randy Stechert's specialty. It wasn't until two springs later, in April 1999, that Jesse Jaycox and Edwin McGowan, another DEC biologist, actually found and properly identified a new den, which I'll refer to as the Sour Hill den, fewer than one hundred feet from Sour Mountain's boundary, while conducting a formal timber rattlesnake assessment of the area for the DEC. It turned out that what Stechert had located was not a den but a nearby basking area 270 feet from the den. In over thirty-five years in the field and scores of discoveries of previously unknown dens to his name, this was the only time that Stechert is known to have misidentified the precise location of a den.

6. Breisch, e-mail to author, August 3, 2005.

7. W. S. Brown informed me in November 2003 that Sour Mountain Realty grudgingly removed its fence two weeks after its April 3, 1999, deadline, under the threat of a contempt of court citation. If Sour Mountain were to pursue its permit process with the DEC in the future, the chances of obtaining the permits would be slim at best. The Sour Hill den with its colony of timber rattlesnakes alone was enough to pose serious problems for the permit process and put it into a state of limbo. But then, in the fall of 1999, a second previously unknown den was found in the area, not adjacent to the Sour Mountain property but deep inside it, right where the company wanted to conduct its mining activities. Randy Stechert filled me in on how the second den came to be found, in a telephone interview on March 16, 2004. During the summer of 1999 Jesse Jaycox and Ed McGowan did a radio tracking study on Sour Mountain's property and the adjacent property containing the Sour Hill den. It was part of a timber rattlesnake assessment they were conducting for the DEC. Stechert and W. S. Brown were also under contract with the DEC to assist in this assessment. By late summer, Jaycox and McGowan began tracking a few of their transmitter-implanted snakes back to the Sour Hill den, but noticed that a few of their other implanted snakes were remaining on the Sour Mountain property. Could this mean another den? Stechert remembers Jaycox's excitement on the telephone one evening in late September. He had found what he was convinced was a new den in the area earlier that day, having tracked two snakes to two small groups of snakes on a ledge way within Sour Mountain's boundary lines. A few days later, Jaycox and McGowan returned to the area with Stechert and Brown, and while investigating the outcropping below the ledge, Jaycox and Stechert discovered the den. It turned out that Jaycox had tracked his snakes to a staging area just above the actual den crevice. A thorough report on this study, authored by McGowan, Jaycox, Brown, and Kerpez, was given to the DEC on April 10, 2000.

8. My understanding of the Ramapo Energy case stems largely from W. S. Brown's 2002 unpublished report entitled "Power Plant Defeated by Rattlesnake: A Case in Herpetological Conservation," and also from numerous conversations and communiqués with Brown between October 10, 2002, and May 17, 2003.

9. For an overview by Brown, Stechert, and others on the negative impact of human disturbance on timber rattlesnakes, see William S. Brown, *Biology, Status, and Management of the Timber Rattlesnake (Crotalus horridus): A Guide for Conservation* (Lawrence, Kansas: Society for the Study of Amphibians and Reptiles 1993), 34–36. Stechert uses the term "spook factor" on page 36.

10. My knowledge of the Galick property purchase stems from a telephone interview with Mark DesMeules on April 5, 2003.

11. The friend who accompanied Mark DesMeules into the Galick farm is Tom Tyning, a well-known New England naturalist, biologist, and author.

12. See Madeleine M. Kunin, *Living a Political Life: One of America's First Woman Governors Tells Her Story* (New York: Vintage Books, 1994), 396.

13. Stechert, telephone interview with author, October 10, 2005.
14. Stechert, telephone interview with author, July 2, 2005.
15. Breisch, telephone interview with author, March 24, 2003.
16. Breisch, telephone interview with author, March 26, 2003.

CONCLUSION (p. 166)
1. R. Stechert, conversation with author during field trip, June 19, 2005.

BIBLIOGRAPHY

Amato, Christopher A., and Robert Rosenthal. "Endangered Species Protection in New York after State v. Sour Mountain Realty, Inc." *New York University Environmental Law Journal* 10 (2001): 117–45.

Bailey, John H. "A Stratified Rock Shelter in Vermont." *Bulletin of the Champlain Valley Archaeological Society* (Fort Ticonderoga, New York) 1 (1940): 1–30.

Behler, John L. "The Great American Snake Hunt: On the Trail of Monster Serpents." *Animal Kingdom* 78 (1975): 21–26.

Brown, William S. *Biology, Status, and Management of the Timber Rattlesnake (Crotalus horridus): A Guide for Conservation*. Lawrence, Kansas: Society for the Study of Amphibians and Reptiles, 1993.

———. "Female Reproductive Ecology in a Northern Population of the Timber Rattlesnake, *Crotalus horridus*." *Herpetologica* 47 (1991): 101–15.

———. "Hidden Life of the Timber Rattler." *National Geographic* 17 (1987): 128–38.

———. "Overwintering Body Temperatures of Timber Rattlesnakes (*Crotalus horridus*) in Northeastern New York." *Journal of Herpetology* 16 (1982): 145–50.

Brown, William S., and David B. Greenberg. "Vertical-tree Ambush Posture in *Crotalus horridus*." *Herpetological Review* 23 (1992): 67.

Brown, William S., Len Jones, and Randy Stechert. "A Case in Herpetological Conservation: Notorious Poacher Convicted of Illegal Trafficking in Timber Rattlesnakes." *Bulletin of the Chicago Herpetological Society* 29 (1994): 74–79.

Brown, W. S., et al. "Movements and Temperature Relationships of Timber Rattlesnakes (*Crotalus horridus*) in Northeastern New York." *Journal of Herpetology* 16 (1982): 151–61.

Burton, Thomas. *Serpent-handling Believers*. Knoxville, Tennessee: University of Tennessee Press, 1993.

Clark, Anne Marie, et al. "Phylogeography of the Timber Rattlesnake (*Crotalus horridus*) based on mtDNA Sequences." *Journal of Herpetology* 37 (2003): 145–54.

Clark, Rulon W. "Diet of the Timber Rattlesnake." *Journal of Herpetology* 36 (2002): 494–99.

Conant, Roger. *A Field Guide to Reptiles and Amphibians*. Boston: Houghton Mifflin, 1958.

Conant, Roger, and Joseph T. Collins. *A Field Guide to Reptiles and Amphibians of Eastern and Central North America*. Third ed. Boston: Houghton Mifflin, 1991.

DeKay, James Ellsworth. *Reptiles and Amphibians*. 1842. This is volume 1 of *Zoology of New York*, which is part 3 of a thirty-volume set under the title, *Natural History of New York*. New York: Appleton and Company, 1842–1894.

DesMeules, Mark. "Vermont's Timber Rattlesnake: Historic Distribution, Current

Status, and Conservation Outlook." In Thomas F. Tyning, editor, *Conservation of the Timber Rattlesnake in the Northeast*. Lincoln, Massachusetts: Massachusetts Audubon Society, 1992.

Ditmars, R. L. "Serpents of the Eastern States." *Bulletin of the New York Zoological Society* 32 (1929): 83–120.

———. *The Reptiles of North America*. Garden City, New York: Doubleday, 1946.

———. *Snakes of the World*. New York, New York: Macmillan, 1931.

Ernst, Carl H. *Venomous Reptiles of North America*. Washington: Smithsonian Institution Press, 1992.

Fiske, Milan. "A Personal Account of Being Bitten by a Rattler." *The Conservationist* (New York Department of Environmental Conservation) 36 (1981): 30–31.

Fitch, Henry S. *A Kansas Snake Community: Composition and Changes Over 50 Years*. Malabar, Florida: Krieger Publishing Company, 1999.

Fitch, H. S., et al., "A Field Study of the Timber Rattlesnake in Leavenworth County, Kansas." *Journal of Kansas Herpetology* 11 (2004): 18–24.

Hardy, David L., Sr. "A Review of First Aid Measures For Pitviper Bite in North America with an Appraisal of Extractor Suction and Stun Gun Electroshock." In Jonathan A. Campbell and Edmund D. Brodie, Jr., editors, *Biology of the Pitvipers*. Tyler, Texas: Selva Press, 1992.

Henry, Alexander. *Travels and Adventures in Canada and the Indian Territories Between the Years 1760 and 1776*. New York, 1809.

Holman, J. Alan. *Pleistocene Amphibians and Reptiles in North America*. New York: Oxford University Press, 1986.

Keyler, Daniel E. "Venomous Snakebites: Minnesota and Upper Mississippi River Valley 1982–2002." *Minnesota Herpetological Society Occasional Paper Number 7* (2005): 1–28.

Klauber, L. M. *Rattlesnakes: Their Habits, Life Histories, and Influence on Mankind*. 2 vols. Berkeley, California: University of California Press, 1956.

Kunin, Madeleine M. *Living A Political Life: One of America's First Woman Governors Tells Her Story*. New York: Vintage Books, 1994.

Martin, William H. "Phenology of the Timber Rattlesnake (*Crotalus horridus*) in an Unglaciated Section of the Appalachian Mountains." In Jonathan A. Campbell and Edmund D. Brodie, Jr., editors, *Biology of the Pitvipers*. Tyler, Texas: Selva Press, 1992.

———. "Life History Constraints on the Timber Rattlesnake (*Crotalus horridus*) at Its Climatic Limits." In Gorden W. Schuett, Mats Höggren, Michael E. Douglas, and Harry W. Greene, editors, *Biology of the Vipers*. Eagle Mountain, Utah: Eagle Mountain Publishing, 2002.

———. "The Timber Rattlesnake in the Northeast: Its Range, Past and Present." *Bulletin of the New York Herpetological Society* 17 (1982): 15–20.

Mathews, Thomas. "Rattlesnake Colonel." *New England Quarterly* 10 (1937): 341–45.

Merrow, Jed S., and Todd Aubertin. "*Crotalus horridus* (Timber Rattlesnake) Reproduction." *Herpetological Review* 36(2005):192.

Norris, Robert. "Venom Poisoning by North American Reptiles." In Jonathan A. Campbell and William W. Lamar. *The Venomous Reptiles of the Western Hemisphere*. Vol. 2. Ithaca, New York: Cornell University Press, 2004.

Palmer, Thomas. *Landscape with Reptile: Rattlesnakes in an Urban World*. New York: Ticknor and Fields, 1992.

Reinert, Howard K., and Robert T. Zappalorti. "Field Observation of the Association of Adult and Neonatal Timber Rattlesnakes, *Crotalus horridus*, with Possible Evidence for Conspecific Trailing." *Copeia* (1988): 1057–59.

Rice, Arthur F. "Vermont's Rattlesnakes." *Forest and Stream* 44 (1895), 389.

Rubio, M. *Rattlesnake; Portrait of a Predator*. Washington: Smithsonian Institution Press, 1998.

Russell, F. E. *Snake Venom Poisoning*. Philadelphia, Pennsylvania: J. B. Lippincott Company, 1980.

Stechert, Richard (Randy). "Distribution and Population Status of *Crotalus horridus* in New York and Northern New Jersey." In T. F. Tyning, editor, *Conservation of the Timber Rattlesnake in the Northeast*. Lincoln, Massachusetts: Massachusetts Audubon Society, 1992.

Stiles, Fred Tracy. *Old Days–Old Ways: More History and Tales of the Adirondack Foothills*. Fort Edward, New York: Washington County Historical Society, 1984.

Tennant, Alan. *Snakes of North America: Eastern and Central Regions*. Latham, Maryland: Lone Star Books, 2003.

Thompson, Zadac. *History of Vermont: Natural, Civil and Statistical, in three parts*. Burlington, Vermont: Chauncey Goodrich, 1842.

INDEX